AQA Science

Exclusively endorsed and approved by AQA

Revision Guide

Nigel English

Series Editor: Lawrie Ryan

GCSE Biology

Nelson Thornes
a Wolters Kluwer business

Published in 2006 by:
Nelson Thornes Ltd
Delta Place
27 Bath Road
CHELTENHAM
GL53 7TH
United Kingdom

06 07 08 09 10 / 10 9 8 7 6 5 4 3 2

A catalogue record for this book is available from the British Library

ISBN 0 7487 8312 1

Cover photographs: snail by Gerry Ellis/Digital Vision LC (NT); embryo by Biophoto/Science Photo Library; *E. coli* bacteria by Dr Gary Gaugler/Science Photo Library
Cover bubble illustration by Andy Parker

Illustrations by Bede Illustration
Page make-up by Wearset Ltd

Printed and bound in Croatia by Zrinski

Acknowledgements

Axon Images 41mr, 49mr; **Corbis V98 (NT)** 10br; **Corel 11 (NT)** 7b; **Corel 18 (NT)** 32m; **Corel 467 (NT)** 36bl; **Corel 511 (NT)** 7t; **Corel 588 (NT)** 8b; **Corel 706 (NT)** 30br; **Corel 765 (NT)** 90br; **David Buffington/Photodisc 67 (NT)** 5tl; **Digital Vision 7 (NT)** 35tl; **Digital Vision 15 (NT)** 21br; **Gerry Ellis/Digital Vision JA (NT)** 22b; **Karl Ammann/Digital Vision AA (NT)** 23t; **Nigel English** 54tr; **Photodisc 10 (NT)** 63tr; **Photodisc 19 (NT)** 41br, 93l; **Photodisc 29 (NT)** 37m; **Photodisc 45 (NT)** 61l, 62br; **Photodisc 50 (NT)** 2b; **Photodisc 54 (NT)** 16br; **Photodisc 67 (NT)** 6b; **Photodisc 71 (NT)** 62bl, 63tl; **Ringwood Brewery** 87tl, 87tm, 89ml, 89bl; **Science Photo Library** 14br, /**Adam Hart-Davis** 24tr, /**Anthony Mercieca** 23b, /**BSIP VEM** 9l, /**BSIP/Beranger** 73br, 82m, /**CNRI** 12m, 12b, 68bl, /**Cordelia Molloy** 15bl, 46b, /**David Hall** 87br, 92br, /**David Nunuk** 25tl, /**Div. of Computer Research and Technology, National Institute of Health** 21mr, 27tl, /**Eye Of Science** 57br, /**J.C. Revy** 43bl, 57tr, /**Kenneth W. Fink** 33r, /**Kent Wood** 15tr, /**Mark Clarke** 57ml, /**Mauro Fermariello** 90tl, /**Maximilian Stock** 91r, /**Renee Lynn** 24b, /**Rosenfeld Image Ltd** 90tr, /**St Mary's Hospital School** 92tr, /**Steve Gschmeissner** 26b, 43br, /**Tony Craddock** 13tl; **Topfoto.co.uk/David R Frazier/The Image Works** 90bm; **Topfoto.co.uk/National Pictures** 11m, /**The Image Works** 8t

Many thanks for the contributions made by Paul Lister, Ann Fullick and Niva Miles.

Picture research by Stuart Sweatmore, Science Photo Library and johnbailey@ntlworld.com.

Every effort has been made to trace all the copyright holders, but if any have been overlooked the publisher will be pleased to make the necessary arrangements at the first opportunity.

How to answer questions

Question speak

Command word or phrase	What am I being asked to do?
compare	State the similarities and the differences between two or more things.
complete	Write words or numbers in the gaps provided.
describe	Use words and/or diagrams to say how something looks or how something happens.
describe, as fully as you can	There will be more than one mark for the question so make sure you write the answer in detail.
draw	Make a drawing to show the important features of something.
draw a bar chart / graph	Use given data to draw a bar chart or plot a graph. For a graph, draw a line of best fit.
explain	Apply reasoning to account for the way something is or why something has happened. It is not enough to list reasons without discussing their relevance.
give / name / state	This only needs a short answer without explanation.
list	Write the information asked for in the form of a list.
predict	Say what you think will happen based on your knowledge and using information you may be given.
sketch	A sketch requires less detail than a drawing but should be clear and concise. A sketch graph does not have to be drawn to scale but it should be the appropriate shape and have labelled axes.
suggest	There may be a variety of acceptable answers rather than one single answer. You may need to apply scientific knowledge and/or principles in an unfamiliar context.
use the information	Your answer **must** be based on information given in some form within the question.
what is meant by	You need to give a definition. You may also need to add some relevant comments.

Diagrams

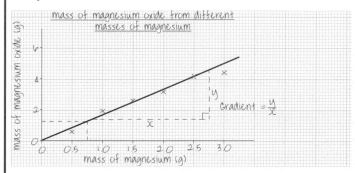

Things to remember:

- Draw diagrams in pencil.
- The diagram needs to be large enough to see any important details.
- Light colouring could be used to improve clarity.
- The diagram should be fully labelled.
- Label lines should be thin and end at the point on the diagram that corresponds to the label.

How long should my answer be?

Things to consider:

1 How many lines have been given for the answer?
- One line suggests a single word or sentence. Several lines suggest a longer and more detailed answer is needed.

2 How many marks is the answer worth?
- There is usually one mark for each valid point. So for example, to get all of the marks available for a three mark question you will have to make three different, valid points.

3 As well as lines, is there also a blank space?
- Does the question require you to draw a diagram as part of your answer?
- You may have the option to draw a diagram as part of your answer.

Graphs

Things to remember:

- Choose sensible scales so the graph takes up most of the grid.
- Don't choose scales that will leave small squares equal to 3 as it is difficult to plot values with sufficient accuracy on such scales.
- Label the axes including units.
- Plot all points accurately by drawing small crosses using a fine pencil.
- Don't try to draw a line through every point. Draw a line of best fit.
- A line of best fit does not have to go through the origin.
- When drawing a line of best fit, don't include any points which obviously don't fit the pattern.
- The graph should have a title stating what it is.
- To find a corresponding value on the y-axis, draw a vertical line from the x-axis to the line on the graph, and a horizontal line across to the y-axis. Find a corresponding value on the x-axis in a similar way.
- The gradient or slope of a line on a graph is the amount it changes on the y-axis divided by the amount it changes on the x-axis. (See the graph above.)

Calculations

- Write down the equation you are going to use, if it is not already given.
- If you need to, rearrange the equation.
- Make sure that the quantities you put into the equation are in the right units. For example you may need to change centimetres to metres or grams to kilograms.
- Show the stages in your working. Even if your answer is wrong you can still gain method marks.
- If you have used a calculator make sure that your answer makes sense. Try doing the calculation in your head with rounded numbers.
- Give a unit with your final answer, if one is not already given.
- Be neat. Write numbers clearly. If the examiner cannot read what you have written your work will not gain credit. It may help to write a few words to explain what you have done.

How to use the 'How Science Works' snake

The snake brings together all of those ideas that you have learned about 'How Science Works'. You can join the snake at different places – an investigation might start an observation, testing might start at trial run.

How do you think you could use the snake on how plants are affected by acid rain? Try working through the snake using this example – then try it on other work you've carried out in class.

Remember there really is no end to the snake – when you reach the tail it is time for fresh observations. Science always builds on itself – theories are constantly improving.

OBSERVATION

I wonder why…

HYPOTHESIS

Perhaps it's because…

PREDICTION

I think that if…

I should be honest and tell it as it is. Does the data support or go against my hypothesis?

Is it a linear (straight line) relationship – positive, negative or directly proportional (starting at the origin)? or is it a curve – complex or predictable?

Which of these should I use?
- Bar chart
- Line graph
- Scatter graph

RELATIONSHIP SHOWN BY DATA

PRESENTING DATA

Am I going further than the data allows me?

Are the links I have found – causal, by association or simply by chance?

Have I given a balanced account of the results?

CONCLUSION

My conclusion would be more reliable and valid if I could find some other research to back up my results.

Just how reliable (trustworthy) was the data? Would it be more reliable if somebody else repeated the investigation? Was the data valid – did it answer the original question?

EVALUATION

USE SECONDARY DATA

There are still many questions that we cannot answer in scienc

Should the variables I use be continuous (any value possible), discrete (whole number values), ordered (described in sequence) or categoric (described by words)?

I will try to keep all other variables constant, so that it is a fair test. That will help to make it valid.

DESIGN

CONTROL VARIABLE

TRIAL RUN

This will help to decide the:
- Values of the variables
- Number of repeats
- Range and interval for the variables

Can I use my prediction to decide on the variable I am going to change (independent) and the one I am going to measure (dependent)?

Are my instruments sensitive enough?

Will the method give me accuracy (i.e. data near the true value)? Will my method give me enough precision and reliability (i.e. data with consistent repeat readings)?

PREPARE A TABLE FOR THE RESULTS

I'll try to keep random errors to a minimum or my results will not be precise. I must be careful!

Are there any systematic errors? Are my results consistently high or low?

CARRY OUT PROCEDURE

Are there any anomalies (data that doesn't follow the pattern)? If so they must be checked to see if they are a possible new observation. If not, the reading must be repeated and discarded if necessary.

I should be careful with this information. This experimenter might have been biased – must check who they worked for; could there be any political reason for them not telling the whole truth? Are they well qualified to make their judgement? Has the experimenter's status influenced the information?

I should be concerned about the ethical, social, economic and environmental issues that might come from this research.

The final decisions should be made by individuals as part of society in general.

Could anyone exploit this scientific knowledge or technological development?

TECHNOLOGICAL DEVELOPMENTS

There are questions that science cannot answer at all – such as 'Should we…?' questions.

B1a | Human biology

Checklist

This spider diagram shows the topics in the unit. You can copy it out and add your notes and questions around it, or cross off each section when you feel confident you know it for your exams.

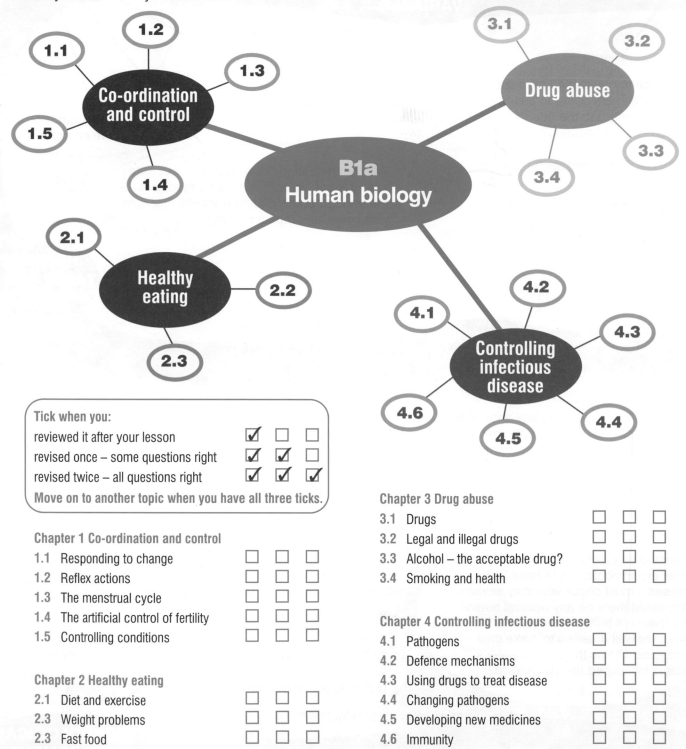

Tick when you:

reviewed it after your lesson	✓	☐	☐
revised once – some questions right	✓	✓	☐
revised twice – all questions right	✓	✓	✓

Move on to another topic when you have all three ticks.

Chapter 1 Co-ordination and control

1.1	Responding to change	☐	☐	☐
1.2	Reflex actions	☐	☐	☐
1.3	The menstrual cycle	☐	☐	☐
1.4	The artificial control of fertility	☐	☐	☐
1.5	Controlling conditions	☐	☐	☐

Chapter 2 Healthy eating

2.1	Diet and exercise	☐	☐	☐
2.3	Weight problems	☐	☐	☐
2.3	Fast food	☐	☐	☐

Chapter 3 Drug abuse

3.1	Drugs	☐	☐	☐
3.2	Legal and illegal drugs	☐	☐	☐
3.3	Alcohol – the acceptable drug?	☐	☐	☐
3.4	Smoking and health	☐	☐	☐

Chapter 4 Controlling infectious disease

4.1	Pathogens	☐	☐	☐
4.2	Defence mechanisms	☐	☐	☐
4.3	Using drugs to treat disease	☐	☐	☐
4.4	Changing pathogens	☐	☐	☐
4.5	Developing new medicines	☐	☐	☐
4.6	Immunity	☐	☐	☐

What are you expected to know?

Chapter 1 Co-ordination and control (See students' book pages 24–35)

- Co-ordination of what happens inside your body, and responses to changes outside your body, rely on hormones and the nervous system.

- The nervous system involves neurones and impulses, which are electrical (except at a synapse).

- A reflex is a rapid, automatic response to a stimulus.

- Internally controlled conditions in your body include:
  - water content
  - ion content
  - temperature
  - blood sugar level.

- Hormones are chemicals produced by glands and carried in your bloodstream.

- The menstrual cycle is controlled by three hormones:
  - follicle stimulating hormone (FSH)
  - oestrogen
  - luteinising hormone (LH).

- These hormones can be used to control a woman's fertility.

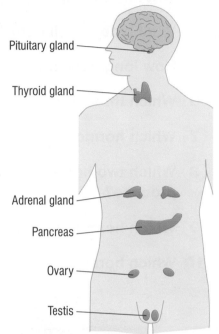

Pituitary gland

Thyroid gland

Adrenal gland

Pancreas

Ovary

Testis

Chapter 2 Healthy eating (See students' book pages 38–45)

- A healthy diet contains the right balance of different foods.

- Too little or too much food can lead to health and weight problems.

- Too much cholesterol or salt can lead to problems in your blood system.

- Cholesterol is carried in the body by low-density lipo-proteins (LDLs are 'bad') or high-density lipo-proteins (HDLs are 'good').

Chapter 3 Drug abuse (See students' book pages 48–59)

- Most drugs harm the body. Even those developed to cure disease can have harmful side effects.

- Drugs can be addictive. If you try to stop taking them you may suffer from withdrawal symptoms.

Chapter 4 Controlling infectious disease (See students' book pages 62–75)

- Microorganisms that cause infection are called 'pathogens'. These include bacteria and viruses.

- The body defends itself by:
  - ingesting pathogens
  - producing antibodies
  - producing antitoxins.

- We can use antibiotics and vaccination to control infection.

- Some pathogens have developed resistance to antibiotics.

1. How are impulses transmitted in the nervous system?

2. Where are hormones produced?

3. What is a synapse?

4. What type of cell detects a change in external conditions, e.g. temperature?

5. How long does the menstrual cycle last?

6. Which three hormones control the cycle?

7. Which hormone stimulates the eggs to be released by an ovary?

8. Which two hormones, controlling the menstrual cycle, does the pituitary gland produce?

9. Which hormone is used in the contraceptive pill?

10. Which hormone is used in the 'fertility' treatment of women?

---

students' book page 24   **B1a 1.1**   # Responding to change

KEY POINT

Control of the body's functions and responses involves hormones (chemicals) and the nervous system (electrical impulses).

The body must respond to internal and external conditions.

- You have glands that produce hormones. The hormones are transported around your body by the blood.
- Electrical impulses pass along the nervous system.
- All responses must be co-ordinated.

Key words: gland, hormone, impulse, nervous system

BUMP UP YOUR GRADE

Remember that impulses are transmitted by chemicals in the hormone system and electrical impulses in the nervous system.

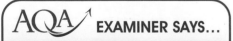

AQA EXAMINER SAYS...

There are often questions about how the nervous and hormone systems are different.

Make sure that you know!

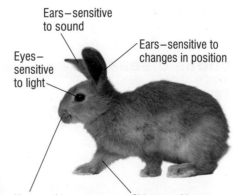

Ears—sensitive to sound

Ears—sensitive to changes in position

Eyes—sensitive to light

Nose and tongue—sensitive to chemicals

Skin—sensitive to touch, pressure, pain and temperature

Being able to detect changes in the environment is important

CHECK YOURSELF

1 How do hormones reach their target organs?

2 What type of organ produces hormones?

3 How are impulses passed along the nervous system?

Reflex actions

Here are the steps involved in a reflex action:

- A receptor detects a stimulus (e.g. sharp pain).
- A sensory neurone transmits the impulse.
- A relay neurone passes the impulse on.
- A motor neurone is stimulated.
- The impulse is sent to the effector (muscle or gland).
- Action is taken.

At the junction between two neurones is a synapse. Chemicals transmit the impulse across the gap.

Key words: effector, receptor, neurone, synapse, stimulus

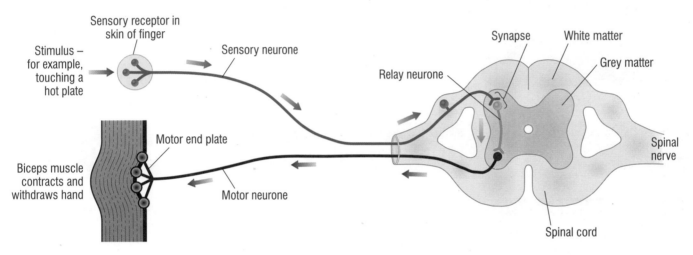

The reflex action which moves your hand away from something hot can save you from a nasty burn!

CHECK YOURSELF

1 How are impulses transmitted across a synapse?

2 What is the function (job) of a relay neurone?

3 What are 'effectors'?

B1a 1.3 The menstrual cycle

KEY POINTS

1 The cycle takes 28 days with ovulation about 14 days into the cycle.
2 The cycle is controlled by three hormones.

- **FSH** is made by the pituitary gland and causes the egg to mature and oestrogen to be produced.
- **Oestrogen** is produced by the ovaries and inhibits the further production of FSH. It stimulates the production of LH and also stimulates the womb lining to develop to receive the fertilised egg.
- **LH** is made by the pituitary gland and stimulates the mature egg to be released.

Key words: stimulate, inhibit, womb, mature

AQA↗ EXAMINER SAYS...

Many students remember the effect each hormone has on egg production but forget how each hormone affects the production of the others. Make sure that you know!

GET IT RIGHT!

Remember what each hormone does during the cycle, especially the effect each one has on the production of other hormones.

CHECK YOURSELF

1 Where is FSH produced?
2 What effect does the production of oestrogen have on the production of FSH and LH?
3 What does LH do?

B1a 1.4 The artificial control of fertility

KEY POINTS

1 The contraceptive pill contains oestrogen that prevents pregnancy.
2 FSH can be given to a woman to help her to produce eggs.

- The contraceptive pill contains oestrogen. This prevents the production of FSH so no eggs mature.
- If a woman cannot produce mature eggs then FSH can be given. This is known as 'fertility treatment'.

Key words: contraception, fertility treatment

Ovary
Ripe egg

1 Fertility drugs are used to make lots of eggs mature at the same time for collection

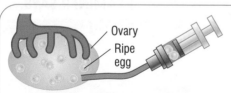

2 The eggs are collected and placed in a special solution in a petri dish

3 A sample of semen is collected

4 The eggs and sperm are mixed in the petri dish

5 The eggs are checked to make sure they have been fertilised and the early embryos are developing properly

6 When the fertilised eggs have formed tiny balls of cells, 1 or 2 of the tiny embryos are placed in the uterus of the mother. Then, if all goes well, at least one baby will grow and develop successfully.

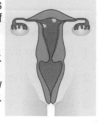

New reproductive technology using hormones and IVF (*in vitro* fertilisation) has helped thousands of infertile couples to have babies

EXAM HINTS

There are issues involved in contraception and fertility treatment. Make sure that you can offer opinions on these issues in an examination. If you only offer a 'one-sided' argument, you may lose up to half of the marks.

CHECK YOURSELF

1 Which hormone is present in the contraceptive pill?
2 Why is FSH used for fertility treatment?

B1a 1.5 Controlling conditions

1 It is very important that the internal conditions of the body are kept within certain limits.
2 Water and ion content, as well as temperature and blood sugar level, are all carefully controlled.

A real help in sport – or a good way of making money? Sports drinks are becoming more and more popular, but do most of us really need them?

Internal conditions that are controlled include:
- water content
- ion content
- temperature
- blood sugar level.

Water is leaving the body all the time as we breathe out and sweat. We lose any excess water in the urine (produced by the kidneys). We also lose ions in our sweat and in the urine.

We must keep our temperature constant otherwise the enzymes in the body will not work properly (or may not work at all).

Sugar in the blood is the energy source for cells. The level of sugar in our blood must be controlled.

Key words: internal conditions, blood sugar, ions

CHECK YOURSELF

1 How do we lose water?

2 Why is it important to control our temperature?

3 Why is sugar in the blood important?

B1a 1 End of chapter questions

1 **Describe briefly the stages of a reflex action.**

2 **What are the differences between the hormone and nervous systems?**

3 **What are the functions (jobs) of FSH, oestrogen and LH?**

4 **Suggest one argument for and one argument against the use of fertility treatment.**

5 **State two internal conditions that need to be controlled.**

6 **Which type of neurone transmits the impulse from a motor to a sensory neurone?**

7 **How does the contraceptive pill work?**

8 **Where is LH produced?**

1. What do we mean by a 'balanced diet'?

2. What does malnourished mean?

3. Give three factors that affect how much energy a person needs.

4. What is meant by 'metabolic rate'?

5. What term is used for people who are very fat?

6. State two diseases linked to being overweight.

7. Where is cholesterol made?

8. What are the two types of lipoproteins that carry cholesterol around the body?

9. Which type of fat increases blood cholesterol level?

10. What condition may result from eating too much salt?

---

students' book page 38

B1a 2.1   Diet and exercise

KEY POINTS

1. A healthy diet is made up from the 'right' balance of the different foods that you need.
2. You are malnourished if you do not have a balanced diet.
3. If you take in more energy than you need you may become fat, and this may result in health problems.

Food provides the energy that your body needs to carry out its activities.

Everyone needs a source of energy to survive – and your energy source is your food. Whatever food you eat – whether you prefer sushi, dahl, or roast chicken – most people eat a varied diet that includes everything you need to keep your body healthy.

AQA EXAMINER SAYS...

The word 'rate' often confuses students in an exam. 'Rate' simply means how fast something is taking place. If you run, the 'rate' at which you use energy increases.

If you exercise you will need more energy.

Exercise increases the metabolic rate. This is the rate at which your body uses energy needed to carry out chemical reactions.

Athletes who spend a lot of time training and playing a sport will have a great deal of muscle tissue on their bodies – up to 40% of their body mass. So they have to eat a lot of food to supply the energy they need.

If it is warm you will need less energy than when it is cold.

The amount of energy you need depends on many things. For example:

- your size
- your sex
- the amount of exercise you do
- outside temperature
- pregnancy.

Key words: energy, metabolic rate, chemical reactions, malnourished

If you live somewhere really cold, you need lots of high-energy fats in your diet. You need the energy to keep warm!

CHECK YOURSELF

1 What do we mean by 'metabolic rate'?

2 State two conditions which would increase the amount of food (energy) that you need.

3 Suggest why exercise increases the metabolic rate.

Weight problems

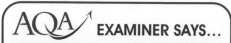
students' book
page 40

KEY POINTS

1 If you eat more food than you need you will put on weight.
2 If you are very fat you are said to be 'obese'.

AQA EXAMINER SAYS...

When asked for a list, many students will remember one or two things. Try to remember three and pick up that extra mark!

If you take in more energy (food) than you need you will become fat.

If you are very fat you are said to be obese.

In spite of some of the media hype, most people are not obese – but the amount of weight people carry certainly varies a great deal!

Obese people are likely to suffer more from:

- arthritis (worn joints)
- diabetes
- high blood pressure
- heart disease.

If you take in less energy (food) than you need you will lose weight.

In developing countries some people have health problems linked to too little food. These include reduced resistance to infection and irregular periods in women.

Key words: obese, arthritis, diabetes, blood pressure, heart disease

Hundreds of thousands of people around the world suffer from the symptoms of malnutrition and starvation. There is simply not enough food for them to eat.

CHECK YOURSELF

1 What word do we use for people who are very fat?

2 What is 'arthritis'?

3 Give three problems, other than arthritis, which obese people may suffer from.

B1a 2.3 Fast food

KEY POINTS

1 'Fast food' is food that has already been prepared. It often contains too much fat and salt.
2 Too much fast food is likely to result in health problems.

GET IT RIGHT!

Make sure that you know which lipoprotein is the good one and which the bad one. It is equally important to know which of the three types of fat will increase the cholesterol level in your body.

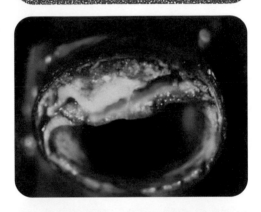

Fast food can contain too much salt and fat.

Too much salt can lead to increased blood pressure.

Cholesterol is made by the liver and the amount made depends upon diet and inherited (genetic) factors. We need cholesterol, but too much in the blood leads to an increased risk of disease of the heart and blood vessels.

Two types of lipoproteins carry cholesterol around the body:

- low-density lipoproteins (LDLs), which are 'bad' and can cause diseases
- high-density lipoproteins (HDLs), which are 'good' for you.

Saturated fat in your diet increases your cholesterol level.

Mono-unsaturated and polyunsaturated fats help reduce cholesterol levels.

We can use statins to stop the liver producing too much cholesterol.

Key words: cholesterol, lipoprotein, saturated, mono-unsaturated, polyunsaturated, statin

CHECK YOURSELF

1 What problem can too much salt in the diet cause?
2 What do lipoproteins do in the body?
3 Which types of fat can help to reduce blood cholesterol levels?

When you get cholesterol building up in the wrong place – like the arteries leading to your heart – it can be very serious indeed

B1a 2 End of chapter questions

1 What can a doctor give you to lower your blood cholesterol level?

2 Which type of fat is likely to increase your blood cholesterol level?

3 If you are fat name three conditions you are more likely to suffer from.

4 What is meant by your 'metabolic rate'?

5 Suggest why you might need more energy on a cold day.

6 In developing countries state one health problem people without enough food might suffer from.

7 What is LDL an abbreviation for?

8 What is 'fast food' likely to contain too much of in terms of a person's diet?

1. What does the word 'illegal' mean?

2. Why are new drugs, developed by scientists, always thoroughly tested before the public can use them?

3. Why was thalidomide used?

4. What are 'withdrawal' symptoms?

5. What do we mean by 'recreational' drugs?

6. Name two illegal drugs.

7. Name two very addictive drugs.

8. What is the addictive substance in tobacco smoke?

9. What effect does carbon monoxide have on the body?

10. What effect does alcohol have on the nervous system?

| students' book page 48 | B1a 3.1 | Drugs |

KEY POINTS

1. All drugs can cause problems whether they are illegal or legal.
2. Many drugs are addictive. If you try to stop taking them this can result in severe withdrawal symptoms.

GET IT RIGHT!

Many problems today are caused by legal drugs, e.g. tobacco and alcohol because they are so widely used (and abused).

Useful drugs, made from natural substances, have been used by indigenous people for a very long time.

When we develop new drugs to help people, we have to test them over a long time to make sure that there are no serious side effects.

- Thalidomide was used as a sleeping pill and to prevent morning sickness in pregnant women. It had very serious side effects on fetuses developing in the womb. It is now used to help cure leprosy.

Millions of pounds worth of illegal drugs are brought into the UK every year. It is a constant battle for the police to find and destroy drugs like these.

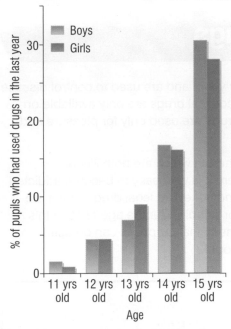

Most of the young people who have used drugs have smoked cannabis – but the number of 15-year-old students who have tried drugs is causing a lot of concern

- Recreational drugs are used by people for pleasure.
- Heroin and cocaine are recreational drugs. They are very addictive and illegal.
- Cannabis is a recreational drug. It is also illegal. It is argued that using cannabis can lead to using 'harder' drugs.
- If you try to stop taking addictive drugs you will suffer withdrawal symptoms.

Key words: illegal, addictive, withdrawal, recreational, indigenous

DON'T LET DRUG DEALERS CHANGE THE FACE OF YOUR NEIGHBOURHOOD.
Call Crimestoppers anonymously on 0800 555 111.

Drugs can seem appealing, exciting and fun when you first take them. Many people use them for a while and then leave them behind. But the risks of addiction are high, and no-one can predict who drugs will affect most.

BUMP UP YOUR GRADE

You may get a question in an examination asking whether taking cannabis leads to taking harder drugs. You must try to make both sides of the argument, as there is no 'right' answer. If you *only* make an argument for or against, you will lose up to half of the marks.

CHECK YOURSELF

1 At the present time, what is thalidomide used to help cure?

2 Name one recreational drug.

3 Suggest why legal recreational drugs are possibly more of a problem than illegal ones.

B1a 3.2 Legal and illegal drugs

EXAMINER SAYS...

Make sure that you know the difference between a drug used as a medicine and a drug that is simply used for pleasure (recreationally). You must also know the difference between legal drugs and illegal drugs.

Medicinal drugs are developed over many years and are used to control disease or help people that are suffering. Many medicinal drugs are only available on prescription from a doctor. Recreational drugs are used only for pleasure and affect the brain and the nervous system.

Recreational drugs include cannabis and heroin which are both illegal. As recreational drugs affect the nervous system it is very easy to become addicted to them. Nicotine and caffeine (in coffee and coke) are legal drugs which are used recreationally, alcohol is also legal for people over the age of 18 in this country. Some drugs which are used for medicinal purposes can be used illegally, e,g, stimulants used by sports people.

CHECK YOURSELF

1 Name one drug used recreationally which is legal.

2 Why can taking recreational drugs lead to addiction?

3 Why do some people say that there is a bigger problem with legal, recreational drugs as opposed to illegal, recreational drugs?

B1a 3.3 Alcohol – the acceptable drug?

- Alcohol slows down the nervous system and therefore slows down your reactions. This will cause problems when driving.
- Too much alcohol leads to loss of self control.
- Drinking too much alcohol may cause a person to lose consciousness or go into a coma.
- The use of alcohol over a long time will damage the liver (cirrhosis) and brain.

Key words: alcohol, coma, liver damage, brain damage

EXAMINER SAYS...

Remember that alcohol is legal. However, it still causes many more problems in our society than illegal drugs such as heroine and cocaine as it is much more widely used.

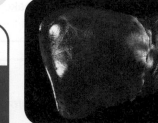

A healthy liver

Your liver deals with all the poisons you put into your body. But if you drink too much alcohol, your liver may not be able to cope. The difference between the healthy liver and the liver with cirrhosis shows just why people are warned against heavy drinking!

Diseased liver from a drinker with cirrhosis

CHECK YOURSELF

1 What effect does alcohol have on the nervous system?

2 Which two organs of the body does alcohol damage?

3 Why should you not 'drink and drive'?

B1a 3.4   Smoking and health

KEY POINTS

1 Smoking tobacco is legal.
2 The use of tobacco causes a range of health problems and will probably lead to an early death.

Cigarette smoking increases your risk of developing many serious and fatal diseases. Every packet of cigarettes sold in the UK has to carry a clear health warning. Yet people still buy them in their millions!

- It is not illegal to smoke tobacco over the age of 16 years.
- The nicotine in tobacco smoke is addictive.
- Tobacco smoke also contains cancer causing chemicals (carcinogens).
- The carbon monoxide in tobacco smoke reduces the amount of oxygen the blood can carry.
- Pregnant women who smoke have babies with lower birth weights, as the babies do not get enough oxygen.

Key words: nicotine, tobacco, carcinogen, carbon monoxide

 EXAMINER SAYS...

Remember it is the nicotine in tobacco smoke that is addictive, and other chemicals in the smoke that cause cancer.

GET IT RIGHT!

Babies born to mothers that smoke have low birth weights. The babies receive less oxygen to release the energy from food (in respiration) that they need to grow properly.

CHECK YOURSELF

1 Why can smoking tobacco cause cancer?

2 What is the addictive substance in cigarettes?

3 Why do the babies born to mothers that smoke have lower birth weights?

B1a 3   End of chapter questions

1 **What effect does alcohol have on the nervous system?**

2 **How does carbon monoxide affect the body?**

3 **Why is it dangerous if you 'drink and drive'?**

4 **Where do the chemicals that cause lung cancer come from?**

5 **Why does it take so long for a new medicinal drug to be made available?**

6 **Give one reason why some people say that taking cannabis can lead to taking 'hard' drugs.**

7 **What is the general name for a substance that may cause cancer?**

8 **Name one legal, recreational drug.**

1. What is a 'pathogen'?

2. Name two organisms that can act as pathogens.

3. What is the other name for the poisons that pathogens produce?

4. What type of cell produces antibodies?

5. What is an anti-toxin?

6. What do painkillers do?

7. What is an antibiotic?

8. Why is it so difficult to kill viruses?

9. Why is MRSA so dangerous?

10. What is in a vaccine?

---

students' book page 62

B1a 4.1 — Pathogens

KEY POINT

Pathogens are microorganisms that cause infectious disease.

EXAM HINTS

Remember in an exam answer that pathogens reproduce before they make enough toxins to make you feel ill.

BUMP UP YOUR GRADE

You will 'bump' your grade in an exam answer from C/D to A by not just knowing *what* Semmelweiss discovered about the transfer of infection, but *why* it took so long for his ideas to be accepted.

- Pathogens cause infectious disease.
- Some bacteria and viruses are pathogens. These bacteria and viruses reproduce inside the body producing poisons (toxins) that make us feel ill.
- Semmelweiss discovered that infection could be transferred between patients in a hospital. He said that washing your hands between treating patients helped stop the transfer of infection. However, it was many years before other doctors took his ideas seriously.

Key words: pathogen, bacteria, virus, toxin, infectious

Ignaz Semmelweiss – his battle to persuade medical staff to wash their hands to prevent infections is still going on today!

CHECK YOURSELF

1 Name the two microorganisms that can cause infection.

2 How did Semmelweiss suggest that you could control the spread of infection in a hospital?

3 Why did it take so long for others to accept Semmelweiss's ideas?

Defence mechanisms

White blood cells do three things to help us protect ourselves:

- They can ingest pathogens. This means that they digest and destroy them.
- They produce antibodies to help destroy particular pathogens.
- They produce antitoxins to counteract (neutralise) the toxins that pathogens produce.

Key words: white blood cell, ingest, antibody, antitoxin

Droplets carrying millions of pathogens fly out of your mouth and nose at up to 100 miles an hour when you sneeze!

EXAMINER SAYS...

In the exam most students will only remember one or, at most, two things that white blood cells do. Try to remember all three!

GET IT RIGHT!

Never write in an exam that the white blood cells eat the pathogens! They ingest them.

CHECK YOURSELF

1 What is the best way to prevent infection?

2 What do antibodies do?

3 How do antitoxins get rid of toxins?

Using drugs to treat disease

- Antibiotics kill infective bacteria in the body. Penicillin is a well-known antibiotic.
- Viruses are much more difficult to kill, as they live inside the cells.
- Painkillers make you feel better, but do nothing to get rid of the disease causing the pain.

Key words: antibiotic, penicillin, painkiller

GET IT RIGHT!

Viruses are harder to get rid of, as they reproduce inside cells. If you destroy the virus you can easily destroy the cell as well.

 EXAMINER SAYS...

Remember that painkillers only relieve the symptoms of a disease.

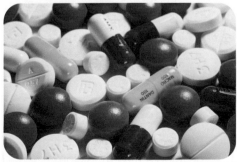

Penicillin was the first antibiotic. Now we have many different types that kill different types of bacteria. In spite of this, scientists are always on the look out for new antibiotics to keep us ahead in the battle against the pathogens.

CHECK YOURSELF

1 What do antibiotics do?

2 Why is it so difficult to get rid of viruses?

3 What is the job of painkillers?

B1a 4.4 — Changing pathogens

If a pathogen changes by mutating or through natural selection, then it will be very difficult to control.

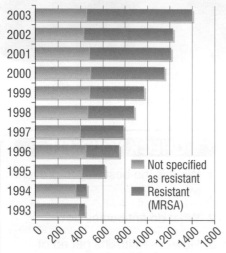

Legend:
- Not specified as resistant
- Resistant (MRSA)

Source: National Statistics Office

The growing impact of MRSA in our hospitals can be seen from this data

The MRSA 'super bug' is a bacterium that has evolved in hospitals through natural selection. It is resistant to commonly used antibiotics.

Some pathogens, particularly viruses, can mutate resulting in new forms.

Very few people are immune to these new pathogens and so disease can spread quickly within a country (epidemic) or across countries (pandemic).

Key words: natural selection, mutation, epidemic, pandemic

AQA EXAMINER SAYS…

Remember that bacteria don't 'want' to develop resistance – they just do!

Make sure you understand how bacteria become resistant to antibiotics through natural selection.

GET IT RIGHT!

Mutation takes place in an instant, whereas natural selection is a gradual process over many years.

CHECK YOURSELF

1 What is a 'pandemic'?

2 Why is MRSA called a 'super bug'?

3 What is meant by 'natural selection'?

B1a 4.5 — Developing new medicines

1 A new medicine must be effective, it must be safe and it must be able to be stored for a period of time.
2 New medicines are tested in laboratories to see if they are toxic and on human volunteers to see if they work.

AQA EXAMINER SAYS…

It is important to test new medicines on animals first to see if they are toxic. Some people are against the use of animals for this. In an examination be prepared to argue the case for and against the use of animals in testing new medicines.

It costs a lot of money to develop a new medicine. It also takes a long time. New medicines must be tested to see if they are toxic (poisonous) and to see if they are effective (cure the disease). This work is carried out in laboratories (on animals) and on human volunteers.

If these tests are not thorough enough then a new medicine may have dangerous side effects. Thalidomide is a medicine that was used widely in the 1950s as a sleeping pill. It also helped to prevent 'morning sickness' in pregnant women. It was not tested thoroughly enough and women started to give birth to babies with severe limb abnormalities. It is now not used with pregnant women but is proving an effective treatment for leprosy.

Key words: toxic, side effects, thalidomide

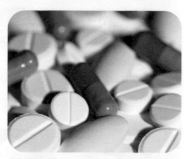

No matter how many medicines we have, there is always room for more as we tackle new diseases!

CHECK YOURSELF

1 Why does it take a long time to develop a new medicine?

2 Why was thalidomide used in the 1950s?

3 Why are some people against the testing of drugs on animals?

B1a 4.6 Immunity

- Dead or inactive forms of an organism can be made into a vaccine. Vaccines can be injected into the body.
- The white blood cells react by producing antibodies. This makes the person immune and prevents further infection, as the body responds quickly by producing more antibodies.
- There is argument over whether some vaccines are completely safe, e.g. the MMR (measles, mumps and rubella) vaccine.

Key words: vaccine, immune

This is how vaccines protect you against dangerous infectious diseases

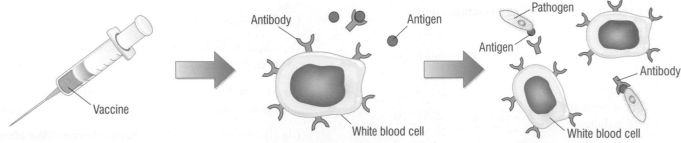

Vaccine

Antibody   Antigen

White blood cell

Pathogen

Antigen

Antibody

White blood cell

Small amounts of dead or inactive pathogen are put into your body, often by injection.

The antigens in the vaccine stimulate your white blood cells into making antibodies. The antibodies destroy the antigens without any risk of you getting the disease.

You are immune to future infections by the pathogen. That's because your body can respond rapidly and make the correct antibody as if you had already had the disease.

GET IT RIGHT!

A vaccine is made from dead or inactive forms of the pathogen.

CHECK YOURSELF

1 What is meant by 'vaccination'?

2 How does the person vaccinated develop immunity?

3 Why are the forms of the pathogen used dead or inactive?

B1a 4   End of chapter questions

1 **What did Semmelweiss find out?**

2 **Why do pathogens make you feel ill?**

3 **What is a vaccine?**

4 **What is an antibody?**

5 **What is meant when we say that white blood cells 'ingest' pathogens?**

6 **Name a commonly used antibiotic.**

7 **Where in the body do viruses live and reproduce?**

8 **What is meant by the word 'mutation'?**

1 The drawing shows a lion.

The lion has organs that contain different receptors. These receptors are sensitive to changes in the environment.

(a) In a lion where are the receptors sensitive to:
(i) light
(ii) sound
(iii) temperature
(iv) balance? (4 marks)

(b) These receptors are part of the nervous system. State two differences between the nervous system and the hormone system. (2 marks)

(c) A lion kills other animals for food. Suggest two adaptations the lion will have, to help it to catch and kill food. (2 marks)

2 Drugs are regularly used by people. There are many different types of drug.

(a) What is meant by a 'recreational' drug? (1 mark)

(b) Name two drugs that are illegal and should not be used by anyone. (2 marks)

(c) Alcohol is legally bought by people over the age of 18 years in the UK.
(i) Which two organs of the body can alcohol damage over a long period of time? (2 marks)
(ii) What effect does alcohol have on the nervous system? (1 mark)

(d) Smoking is legal in the UK over a certain age.
(i) What is the 'addictive' substance in tobacco? (1 mark)
(ii) Name two diseases associated with smoking tobacco. (2 marks)
(iii) When people try to stop smoking they are likely to suffer from 'withdrawal symptoms'. What is meant by the phrase 'withdrawal symptoms'? (1 mark)
(iv) Why is the baby born to a mother who smokes likely to be born underweight? (4 marks)

3 Many children are vaccinated against measles, mumps and rubella (MMR).

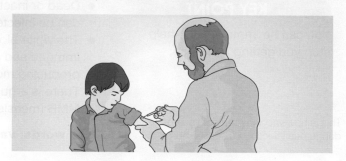

(a) Why does a virus make us feel ill? (1 mark)

(b) Explain, as fully as you can, how a vaccine helps to protect us against disease. (3 marks)

(c) Why is it more difficult to vaccinate successfully against viruses? (2 marks)

(d) (i) What discovery did Semmelweiss make about disease? (2 marks)
(ii) Why did it take so long for his theory to be accepted? (3 marks)

4 It is very important that we eat a healthy diet. More people are eating 'fast food'. If they eat too much, it would make their diet unhealthy.

Fruit   Fast food

(a) What is meant by a 'balanced' diet? (1 mark)

(b) State three factors which will affect how much energy we need. (3 marks)

(c) Fast food can contain too much salt and if you eat too much it may make you fat.
(i) What effect does too much salt have on your health? (1 mark)
(ii) State three conditions that obese (very fat) people are more likely to suffer from. (3 marks)

(d) Cholesterol is found in the body.
(i) Where is cholesterol made? (1 mark)
(ii) If we have too much cholesterol what conditions might this cause? (2 marks)
(iii) How can cholesterol be reduced in the body? (3 marks)

 Test & Assessment Interactive quizzes, answers and hints online!

The answer is worth 4 marks out of the 7 available. The responses worth a mark are underlined in red.

We can improve the answer in several ways:

Write an account of the hormonal control of the menstrual cycle. You should refer in your description to FSH, oestrogen and LH. *(7 marks)*

FSH is made by the pituitary gland and gets to the ovary. It helps the egg in the ovary to mature. Oestrogen is also produced by a gland and causes the lining of the womb to develop and receive a fertilised egg. LH (luteinising hormone) is produced by another gland and causes the ovary to release the egg. All of the three hormones work together and each one affects another. The cycle lasts 28 days when it is regular.

Oestrogen is made by the **ovaries** (a gland is too vague) and **inhibits more FSH** being produced.

LH is produced by the **pituitary gland**.

Stating that all of the hormones affect each other is again too vague.

Stating how long the cycle lasts is not relevant to the question.

The answer is worth 3 marks out of the 5 available. The responses worth a mark are underlined in red.

We can improve the answer in several ways:

If pathogens get into our body they will make us ill. Most pathogens are stopped from entering the body. If some do get in, then the white blood cells help to get rid of them.

1 (a) State two ways that the body can stop pathogens entering the body. *(2 marks)*

   (b) Describe two ways that the white blood cells help get rid of the pathogens. *(3 marks)*

1 (a) 1. If we get a cut then it seals with a scab.
       2. We trap the germs.
   (b) The white cells can digest the bacteria and kill them so that they are destroyed. The white cells also produce something called antibodies. They also produce something to help get rid of the poisons that the pathogens produce which make us feel bad.

Trapping germs is too vague, they are trapped in **mucus** in the nose / airways. **Acid killing pathogens** that get to the stomach is also a good answer.

The fact that antibodies **help the white cells to destroy the pathogens** is worth a mark.

White blood cells do produce chemicals to 'get rid of the poisons'. These are antitoxins but are not relevant to the question, which asks how the white blood cells get rid of the pathogens.

B1b | Evolution and environment

Checklist

This spider diagram shows the topics in the unit. You can copy it out and add your notes and questions around it, or cross off each section when you feel confident you know it for your exams.

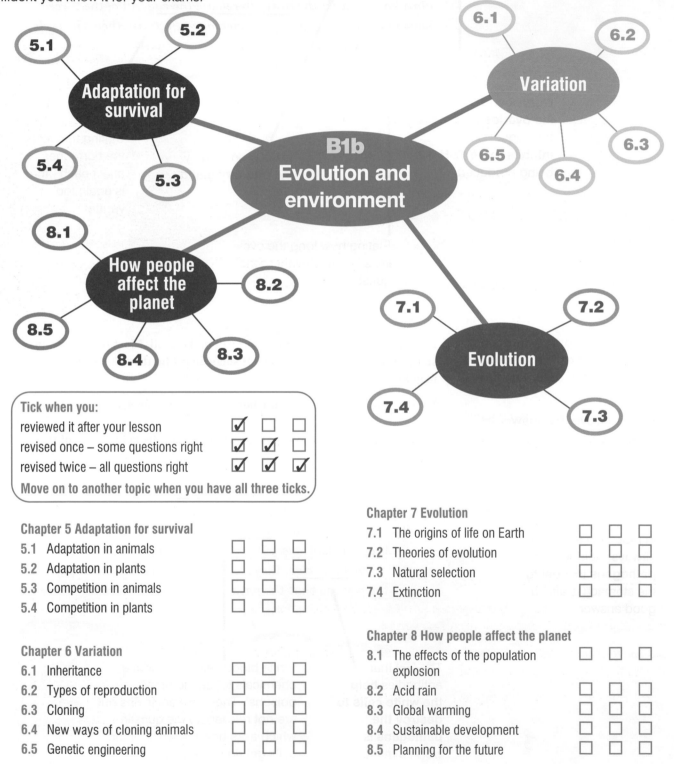

Tick when you:

reviewed it after your lesson	☑	☐	☐
revised once – some questions right	☑	☑	☐
revised twice – all questions right	☑	☑	☑

Move on to another topic when you have all three ticks.

Chapter 5 Adaptation for survival

5.1	Adaptation in animals	☐	☐	☐
5.2	Adaptation in plants	☐	☐	☐
5.3	Competition in animals	☐	☐	☐
5.4	Competition in plants	☐	☐	☐

Chapter 6 Variation

6.1	Inheritance	☐	☐	☐
6.2	Types of reproduction	☐	☐	☐
6.3	Cloning	☐	☐	☐
6.4	New ways of cloning animals	☐	☐	☐
6.5	Genetic engineering	☐	☐	☐

Chapter 7 Evolution

7.1	The origins of life on Earth	☐	☐	☐
7.2	Theories of evolution	☐	☐	☐
7.3	Natural selection	☐	☐	☐
7.4	Extinction	☐	☐	☐

Chapter 8 How people affect the planet

8.1	The effects of the population explosion	☐	☐	☐
8.2	Acid rain	☐	☐	☐
8.3	Global warming	☐	☐	☐
8.4	Sustainable development	☐	☐	☐
8.5	Planning for the future	☐	☐	☐

What are you expected to know?

Chapter 5 Adaptation for survival (See students' book pages 82–91)

- Organisms (animals and plants) all compete for available resources, e.g. food.

- Animals also compete with each other for mates.

- Animals and plants have adaptations so that they can successfully survive in the areas they live in.

Chapter 6 Variation (See students' book pages 94–105)

- Genetic information from parents is passed on to their offspring.

- This information is carried in the genes present in the sex cells (gametes).

- Different genes control different characteristics.

- There are two types of reproduction – sexual and asexual.

- Various techniques can be used to produce new, identical plants. These include tissue culture and taking cuttings. This is known as 'cloning' and is a type of asexual reproduction.

- Cloning is more difficult with animals, but techniques have been developed including embryo transplants, fusion and adult cell cloning.

- Genetic engineering is now widely used to produce organisms with characteristics that we want.

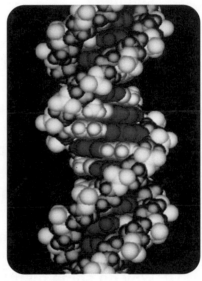

DNA

- There is a lot of argument over the issues of cloning in animals and the genetic engineering of animals and plants.

Chapter 7 Evolution (See students' book pages 108–117)

- Darwin's theory of evolution is now widely accepted (though not by everyone).

- Darwin's theory is based upon variation between members of the same species and the process of natural selection.

- The evidence supporting Darwin's theory includes the fossil record.

- Organisms do become extinct and there are a number of reasons why this happens.

Chapter 8 How people affect the planet (See students' book pages 120–131)

- The human population is growing.

- We are producing much more waste and pollution, and we are also using up the Earth's valuable resources.

- Global warming and the formation of acid rain are two important effects that we are having on the planet.

- There is a lot of effort now being put into 'sustainable development'.

Air pollution

1. State three things that plants compete for.

2. State one other thing that animals compete for.

3. In the Arctic, why might a predator have a white coat in the winter?

4. In the Arctic, why might the prey of a predator have a brown coat in the summer?

5. When answering questions about animals, what do we mean by the 'surface area to volume ratio'?

6. What conditions, in a desert, are likely to cause problems for animals or plants?

7. Why might animals or plants have bright colours?

8. Why do some animals need a territory?

9. In woodland, what might plants be competing for?

10. Why do some plants spread their seed as far away from themselves as possible?

---

students' book page 82

B1b 5.1   Adaptation in animals

KEY POINT

If animals were not adapted to survive in the areas they live in, they would die.

- Animals in cold climates (e.g. in the Arctic) have thick fur and fat under the skin (blubber) to conserve heat.

The Arctic is a cold and bleak environment. However, the animals that live there are well adapted for survival. Notice the large size, small ears, thick coat and white camouflage of this polar bear.

- Some animals in the Arctic (e.g. Arctic fox, Arctic hare) are white in the winter and brown in the summer. This means that they are camouflaged so they are not easily seen.

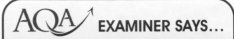
- Bigger animals have smaller surface areas compared to their volume, this means that they can conserve heat more easily but it is also more difficult to lose heat.

Elephants are pretty big but they live in hot, dry climates. Its huge wrinkled skin would cover an animal which was much bigger still. The wrinkles increase the surface area to aid heat loss.

- In dry conditions (desert) animals are adapted to conserve water and to stop them getting too hot. Animals in the desert may hunt or feed at night so that they remain cool during the day.

Animals like this fennec fox have many adaptations to help them cope with the hot dry conditions. How many can you spot?

Key words: adaptation, conserve, surface area : volume ratio, camouflage

CHECK YOURSELF

1 An elephant has a small surface area compared to its volume. How will this affect it in hot climates?

2 Why do very small animals find it difficult to live in the Arctic?

3 Why are Arctic foxes white in the winter?

B1b 5.2 Adaptation in plants

KEY POINTS

1 Plants must be adapted to live in a variety of climates, including deserts.
2 Plants must also be adapted to survive being eaten by animals.

EXAM HINTS

There are many ways plants conserve water in dry environments. These include extensive roots, waxy leaves, small leaves and water storage in stems. Don't limit yourself to one idea in an exam!

- Plants compete for light, water and nutrients.
- In dry conditions, e.g. deserts, plants have become very well adapted to conserve water, e.g. cacti.
- Plants are eaten by animals. Some plants have developed thorns, poisonous chemicals and warning colours to put animals off.

Key words: conserve, warning

CHECK YOURSELF

1 Where might small plants find it difficult to receive enough light?

2 State three possible ways a plant might conserve water.

3 How do animals know not to eat certain plants?

Cacti are well adapted to survive in desert conditions

B1b 5.3 Competition in animals

KEY POINTS

1 Animals compete with each other in many different ways.
2 The most successful ones survive and pass on their genes to the next generation.

 **EXAMINER SAYS...**

There are many ways animals are adapted. Colour (to attract females or as camouflage) and speed to catch food or escape from predators are only two of them.

- Animals compete with each other for water, food, space, mates and breeding sites.
- An animal's territory will be large enough to find water, food and have space for breeding.
- Predators compete with their prey, as they want to eat them.
- Predators and prey may be camouflaged, so that they are less easy to see.
- Prey animals compete with each other to escape from the predators and to find food for themselves.
- Some animals, e.g. caterpillars, may be poisonous and have warning colours so that they are not eaten.

Key words: predator, prey, camouflage

Cheetahs are highly adapted for speed

CHECK YOURSELF

1 Why do animals need a territory?

2 Why do warning colours prevent some animals being eaten?

3 How are giraffes adapted to survive?

B1b 5.4 Competition in plants

KEY POINTS

1 Plants compete for a number of resources.
2 Successful plants have structures and habits that allow healthy growth.

- All plants compete for water, nutrients and light.
  For example, in woodland some smaller plants (e.g. snowdrops) flower before the trees are in leaf, so that they have enough light, water and nutrients.

- Some plants spread their seeds over a wide area so that they do not compete with themselves.

Key words: nutrients, habits

Coconuts will float for weeks, or even months, on ocean currents which can carry them hundreds of miles from their parents – and any other coconuts!

AQA EXAMINER SAYS...

Plants do have structures to enable them to compete, e.g. extensive root systems. However, in an exam question, it is also important to mention that they may have successful growing habits, e.g. they grow quickly so that they gain as much light as possible, or they grow at a time when other plants are dormant, e.g. snowdrops in a wood.

CHECK YOURSELF

1 What three resources do plants compete for?

2 Why do some plants grow and flower early in a wood?

3 Why would plants want to scatter their seeds away from themselves?

B1b 5 End of chapter questions

1 How will an animal learn not to eat certain plants?

2 What will members of the same animal species, e.g. tigers, compete for?

3 Why might it be necessary for some predators to be camouflaged?

4 What conditions in a desert are likely to be difficult for plants and animals?

5 Why do whales have a thick layer of blubber under the skin?

6 State one way in which a cactus is adapted to conserve water.

7 Suggest two reasons why some animals in the desert bury themselves in the sand during the day.

8 Suggest one advantage of a pride of lions causing a herd of animals to run before deciding which one to attack.

1. What are gametes?

2. Where in a cell are chromosomes found?

3. What are chromosomes made up from?

4. Why does sexual reproduction lead to variation in the offspring?

5. How is asexual reproduction different to sexual reproduction?

6. Why are cuttings taken from a plant the same as the parent plant?

7. How do you grow new plants using tissue culture?

8. What does 'cloning' mean?

9. In genetic engineering, how are genes 'cut out' of the chromosome?

10. Why are there arguments about whether genetic engineering should be allowed?

---

students' book page 94   **B1b 6.1**   # Inheritance

KEY POINT

Information from parents is passed to the offspring in the genes.

- The cell nucleus contains chromosomes.

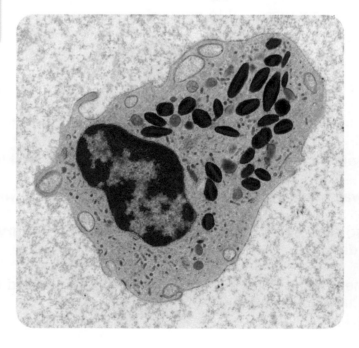

This micrograph shows a highly magnified human cell. In fact the nucleus of the cell would only measure about 0.005 mm! All the instructions for making you and keeping you going are inside this microscopic package. It seems amazing that they work!

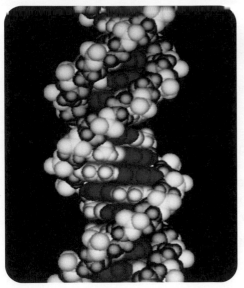

- Chromosomes are made up of genes.
- The male and female sex cells (gametes) contain the genes so the genetic information is passed on to the offspring.
- Genes control the development of the characteristics of the offspring.

Key words: nucleus, chromosome, gene, gamete

DNA! This huge molecule is actually made up of lots of smaller molecules joined together. Each gene is a small section of the big DNA strand.

CHECK YOURSELF

1 What do genes control?

2 How is genetic information passed from parents to offspring?

3 In which part of a cell are chromosomes found?

B1b 6.2 Types of reproduction

KEY POINTS

1 There are two types of reproduction – sexual and asexual.
2 There are very important differences between sexual and asexual reproduction.

AQA EXAMINER SAYS...

Many students forget why one type of reproduction results in variation and the other doesn't. It is simply that in sexual reproduction genes are mixed from the two parents, whereas in asexual reproduction there is only one parent so mixing is impossible!

- Sexual reproduction involves the fusion of sex cells (gametes). There is a mixing of genetic information so the offspring show variation.
- Asexual reproduction does not involve the fusion of sex cells. All of the genetic information comes from one parent. All of the offspring are identical to the parent.
- These identical individuals are known as clones.

Key words: sexual, asexual, clone, variation

  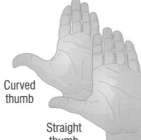  

| Unattached ear lobe | Attached ear lobe | Curved thumb / Straight thumb | Dimples | No dimples |

These are all human characteristics which are controlled by a single pair of genes, so they can be very useful in helping us to understand how sexual reproduction introduces variety and how inheritance works

CHECK YOURSELF

1 Which type of reproduction results in variation?

2 Why does this variation arise?

3 What are identical offspring from one parent called?

B1b 6.3 Cloning

Clones are identical to the parent. Cloning is used to produce new individuals that you want.

- In plants the process is cheap and effective. Plants can be cloned by taking cuttings and growing them, or taking groups of cells and growing them under special conditions (tissue culture).
- With animals it is much more difficult to clone. Embryo transplants are used to clone animals. In this process, embryos are split into smaller groups of cells then each group is allowed to develop in a host animal.

Key words: clone, cutting, tissue culture, embryo transplant, host

CHECK YOURSELF

1 What two types of cloning are commonly used for plants?

2 When using embryo transplants the new individuals are all the same. Why are they all genetically identical?

3 What do we mean by a 'host' animal?

B1b 6.4 New ways of cloning animals

- Fusion cell and adult cell cloning are also used to clone animals.
- In adult cell cloning, the nucleus of the animal you want is placed in an empty cell. This cell is then developed in a different animal.

Key words: fusion cell cloning, adult cell cloning

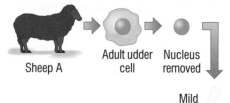

Sheep A → Adult udder cell → Nucleus removed

Mild electric shock

Sheep B → Mature egg cell → Empty egg cell

Nucleus from sheep A fuses with empty egg from sheep B and starts to divide to form an embryo → The cloned embryo is implanted into the uterus of sheep C → Lamb born is clone of sheep A

Adult cell or reproductive cloning is still a very difficult technique – but it holds out the promise of many benefits in the future.

CHECK YOURSELF

1 Look at the diagram above. What type of cell does the nucleus in the fused cell come from?

2 What type of cell provides the empty cell in the process above?

3 What is done to the fused cell to start off the process of cell division in adult cell cloning?

B1b 6.5 Genetic engineering

Genetic engineering involves changing the genetic make-up of an organism.

Genes are 'cut out' of the chromosome of an organism using an enzyme. The genes are then placed in the chromosome of another organism.

The genes may be placed in an organism of the same species so that it has 'desired' characteristics or in a different species. For example, the gene to produce insulin in humans can be placed in bacteria so that they produce insulin.

Many people argue about whether or not genetic engineering is 'right'. Will it create new organisms that we know nothing about? Is it going against nature?

Key words: enzyme, gene, desired

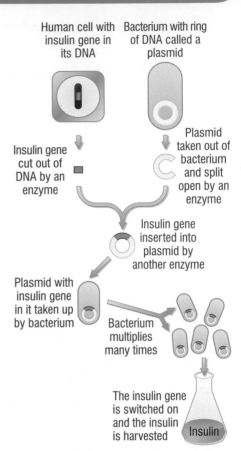

The principles of genetic engineering. A bacterial cell receives a gene from a human being.

B1b 6 End of chapter questions

1 **Why is cloning plants important?**

2 **Why does sexual reproduction result in variation?**

3 **There are three methods of cloning animals. What are they?**

4 **What is 'tissue culture'?**

5 **What is present in the sex cells that result in the characteristics of the parents being passed on to the offspring?**

6 **Why does asexual reproduction result in offspring showing no variation?**

7 **If a grower wants to grow lots of plants that are identical, should he/she use cuttings or tissue culture? Explain your answer.**

8 **Some bacteria have been genetically engineered to produce a human hormone. What is the name of the hormone?**

1. What does the term 'natural selection' mean?

2. What was Lamarck's theory of evolution?

3. Why do most people not believe in Lamarck's theory now?

4. Why did it take so long for Darwin's theory to be accepted?

5. How do fossils provide evidence supporting Darwin's theory of evolution?

6. How long ago did life-forms appear on Earth?

7. In Darwin's theory of evolution, which organisms in a species are the ones that breed?

8. How do these successful members of a species pass on their characteristics to the next generation?

9. What is meant by the word 'extinction'?

10. What is meant by a 'mutation'?

---

students' book
page 108

B1b 7.1   The origins of life on Earth

KEY POINTS

1. We are unsure of when life began on Earth. No one was about!
2. However fossils help us to decide when life began, although it is proving very difficult to find any really good evidence.

AQA↗ EXAMINER SAYS...

Remember in an exam question to link fossils with rocks. They are found in different layers. This means we can put an approximate date on when different animals and plants existed.

It is believed that the Earth is about 4500 million years old and that life began about 3500 million years ago.

There is some debate as to whether the first life developed due to the conditions on Earth, or whether simple life-forms arrived from another planet.

We can date rocks. Fossils are found in rocks, so we can date when different organisms existed.

Key words: fossils, evidence

CHECK YOURSELF

1. When do we think that the first life appeared on our planet?
2. Why can we not be sure where this life came from?
3. How are fossils dated?

This amazing fossil shows two dinosaurs – prehistoric animals which died out millions of years before we appeared on Earth. Fossils can only give us a brief glimpse into the past. We will never know exactly what disaster snuffed out the life of this spectacular reptile all those years ago.

KEY POINTS

1 There are two main theories of evolution. They are those of Lamarck and Darwin.
2 Much evidence now points to the theory of Darwin being the correct one.

AQA EXAMINER SAYS...

In exams many students forget that the members of a species that survive *go on to breed*. This means their genes are passed on. Make sure that you don't forget to write this and therefore miss out on a couple of marks.

Lamarck's theory stated that acquired characteristics can be passed on to the next generation. People found this difficult to believe. For example, if two parents were to build up their muscles in the gym, Lamarck's theory would predict that this characteristic would be passed on to their offspring!

Darwin's theory stated that small changes took place over time. All organisms vary and therefore some are more likely to survive (natural selection). Those that are best adapted breed and pass on their characteristics.

It took a long time for Darwin's theory to be accepted. Many people wanted to believe that God was responsible for the creation of new species.

Key words: acquired, natural selection

In Lamarck's model of evolution, giraffes have long necks because each generation stretched up to reach the highest leaves. So each new generation had a slightly longer neck!

CHECK YOURSELF

1 What did Lamarck mean by an 'acquired characteristic'?

2 Why is it that only some organisms of a species survive to breed?

3 Why did it take so long for people to believe in Darwin's theory?

Natural selection

AQA↗ EXAMINER SAYS…

Remember that organisms are adapted in many different ways. Survival of the fittest does not mean just the fastest. It might mean those able to find food best, fight off disease, survive a 'cold snap' in the weather, etc.

Due to sexual reproduction, there is variation between members of a species. For example, all antelope are different to each other. Those members of a species with the 'best' characteristics survive to breed.

Weaker members of the species may die from:

- disease
- lack of food (or being caught by predators) or
- variation in the climate (a very wet or very cold / hot period of weather).

The survival of organisms with the 'best' characteristics is known as 'survival of the fittest'.

The fact that the best adapted animals and plants survive is known as 'natural selection'.

Key words: characteristics, 'fittest'

Rabbits with the best all-round eyesight, the sharpest hearing and the longest legs will be the ones most likely to escape being eaten by a fox

CHECK YOURSELF

1 Why is there variation between members of the same species?

2 What is meant by the phrase 'survival of the fittest'?

3 What is meant by 'natural selection'?

B1b 7.4 Extinction

KEY POINTS

1 Many species have evolved and then become extinct in the life of the Earth.
2 'Extinction' means that all of the species has been wiped out.
3 There are a number of changes that can cause a species to become extinct.

AQA EXAMINER SAYS...

Make sure when answering a question about extinction that you always remember to use the word 'change'. If there is no change, then species do not become extinct.

CHECK YOURSELF

1 What does the word 'extinct' mean?

2 How might a new competitor cause the extinction of a species?

3 Give two ways that people have caused the extinction of some species.

'Extinction' means that a species that once existed has been completely wiped out.

Extinction can be caused by a number of factors, but always involves a change in circumstances:

● A new disease may kill all members of a species.
● Climate change may make it too cold or hot, or wet or dry, for a species and reduce its food supply.
● A new predator may evolve or be introduced to an area that effectively kills and eats all of the species.
● A new competitor may evolve or be introduced into an area. The original species may be left with too little to eat.
● The habitat the species lives in may be destroyed.

Key words: change, disease, predator, competitor, climate change

The dinosaurs ruled the Earth for millions of years, but when the whole environment changed, they could not adapt and died out. By the time things began to warm up again, mammals, which could control their own body temperature, were becoming dominant. The age of reptiles was over.

B1b 7   End of chapter questions

1 **What is Lamarck's theory as to how new species evolve?**

2 **What, briefly, are the two theories about how life started on Earth?**

3 **Give two reasons why a species could become extinct.**

4 **How do we know when different species lived on the Earth?**

5 **How old do we believe the Earth to be?**

6 **What is meant by the term 'survival of the fittest'?**

7 **How might a new predator arrive on an island?**

8 **Suggest two environmental conditions in an area that, if they changed, could cause problems for the animals that live there.**

B1b 8 Pre Test: How people affect the planet

1. **What is meant by the term 'non-renewable'?**
2. **Give two ways that humans reduce the land available for animals.**
3. **How do humans pollute water supplies?**
4. **What is a fertiliser?**
5. **Which gas is the main cause of acid rain?**
6. **What is a pesticide?**
7. **What organisms are used as water pollution indicators?**
8. **Give two gases that contribute to the 'greenhouse effect'.**
9. **How do forests help to reduce the carbon dioxide level in the atmosphere?**
10. **What do we mean by 'sustainable development'?**

students' book page 120

B1b 8.1 The effects of the population explosion

KEY POINT

The human population is increasing rapidly. So we use up more resources and produce more waste and pollution.

There are increasing numbers of people on our planet.

Many people want and demand a better standard of living.

We are using up raw materials and those that are non-renewable cannot be replaced.

We are producing more waste and the pollution that goes with it.

We are also using land that animals and plants need to live on. It is being used for building, quarrying, farming and dumping waste.

We pollute:

- the water with sewage, fertiliser and toxic chemicals
- the air with gases such as sulfur dioxide and with smoke
- the land with pesticides and herbicides and these can then be washed into the water.

The Earth from space. As the human population of the Earth grows, our impact on the planet gets bigger every day.

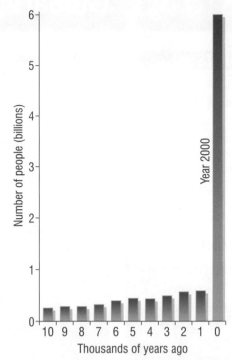

This record of human population growth shows the massive increase during the last few hundred years

CHECK YOURSELF

1 What are the four ways that humans use up land?

2 What are the two chemicals we pollute the land with called?

3 What word do we use to describe raw materials that cannot be replaced?

B1b 8.2 Acid rain

KEY POINTS

1 Sulfur dioxide is mostly responsible for acid rain.
2 Clouds blow across countries, so those producing acid rain often cause problems for somebody else!

- Burning fuels can produce sulfur dioxide (and nitrogen oxides).
- These dissolve in water in the air, forming acidic solutions.
- The solutions then fall as acid rain – sometimes a long way from where the gases were produced.
- Acid rain kills organisms. Enzymes, which control reactions, are very sensitive to pH (acidity or alkalinity).

Key words: acid, dissolve, enzymes

GET IT RIGHT!

Remember the acidity does not dissolve the organism. Living things are killed as their *enzymes* only work in a narrow range of acidity / alkalinity (pH).

CHECK YOURSELF

1 What are the main gases that cause acid rain?

2 Why does acid rain often fall a great distance from where it was produced?

3 Why do some organisms die because of acid rain?

B1b 8.3 Global warming

- Burning fuels (combustion) releases carbon dioxide.
- Cows and rice fields release methane gas.
- Both of these gases are 'greenhouse gases'. As these gases increase in the atmosphere, it retains more heat from the Sun. The Earth is therefore warming up.
- This warming may cause a number of changes in the Earth's climate and cause sea levels to rise.
- We are cutting down forests (deforestation). This is making the problem worse because trees take up carbon dioxide (during photosynthesis). When the trees die, they release this carbon dioxide back to the atmosphere.

AQA EXAMINER SAYS...

Remember methane is also a greenhouse gas. Many students in exam answers only mention carbon dioxide. However, never mention the ozone layer – it has nothing to do with it!

Many scientists believe that this simple warming effect could, if it is not controlled, change life on Earth as we know it

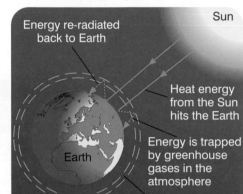

Energy re-radiated back to Earth

Sun

Heat energy from the Sun hits the Earth

Earth

Energy is trapped by greenhouse gases in the atmosphere

Increased levels of CO_2 and methane

Key words: combustion, carbon dioxide, methane, deforestation

CHECK YOURSELF

1 Where is methane produced?

2 Why does more methane and carbon dioxide in the atmosphere mean that it will heat up?

3 Why does deforestation make the problem worse?

B1b 8.4 Sustainable development

As our population rises, we use up more of the Earth's resources. There are many examples, e.g. land, fossil fuels and minerals.

'Sustainable development' means finding ways of reducing this need for more resources.

This may mean finding alternatives to some resources. For example:

- fuels for cars
- recycling what we already have, e.g. plastics, aluminium in cans
- using land that has already been used previously, e.g. for building new homes.

Key words: sustainable, recycle, alternatives

Sustainable woodlands have become an important and attractive part of sustainable development in the UK

AQA EXAMINER SAYS...

There are many examples here. In an exam question, you will be asked to express your views on sustainable development and why it is necessary. You will be able to use any examples that you can think of or are interested in. Make sure that you have a point of view!

CHECK YOURSELF

1 Why will we need to find alternative fuels for cars, lorries, and aeroplanes?

2 Why is it important that we re-use land?

3 Why do you think we should recycle aluminium cans?

B1b 8.5 Planning for the future

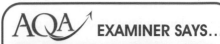

EXAMINER SAYS...

Make sure that you remember the two indicators of pollution. Lichens for air pollution and freshwater invertebrates for water pollution.

- Lichens indicate the level of air pollution. The more species of lichen growing, the cleaner the air.
  Freshwater invertebrates indicate the level of water pollution in the same way. The wider the range of these invertebrates the cleaner the water in the streams, river or pond. Some freshwater invertebrates will only live in polluted water.

- The world population is growing and needs suitable housing. It is important not to use up green areas of the countryside too much but to use areas that have already been built upon (called brown field sites).
  Many countries put aside areas of land which are important for wildlife and will not allow any development on them.

Key words: lichen, freshwater invertebrate, pollution, sustainable

Lichens grow well where the air is clean. In a polluted area there would be far fewer species of lichen growing. This is why they are useful bio-indicators.

CHECK YOURSELF

1 Why is it important to use land previously built upon for new housing?

2 Why is it helpful that some freshwater invertebrates only live in polluted water?

3 Would you expect to find a smaller or wider range of freshwater invertebrates in clean water?

B1b 8 End of chapter questions

1 Give two ways humans are using up more land.

2 Which organism is used as an air pollution indicator?

3 Which chemicals, in organisms, are very pH-sensitive?

4 If a tree is cut down, how might it release its carbon back to the atmosphere?

5 State two ways that humans pollute water.

6 Which gas is the main cause of the 'greenhouse effect'?

7 What are herbicides used for?

8 Name two substances that are now commonly recycled.

1 Baby leopards look like their parents. This is because information is passed on to them when the sex cells of their parents fuse together.

Complete the sentences by using the correct words from the box.

body cells   chromosomes   gametes   genes
membranes   nucleus   sex

Genetic information is passed on to the young leopards in the . . .

Each individual characteristic is controlled by . . .

The thread-like structures that carry information for many characteristics are called . . . .

The part of the cell that carries this genetic information is called the . . . .   (4 marks)

2 Humans affect the environment in a number of different ways.

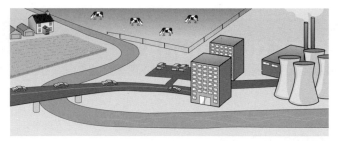

(a) Write down two ways in which humans have reduced the amount of land available for wildlife.   (2 marks)

(b) The farmer may use the following types of chemical. What would he/she use each one for?
   (i)   fertiliser
   (ii) herbicide
   (iii) pesticide   (3 marks)

(c) How might any of these chemicals get into ponds and rivers and pollute them?   (2 marks)

(d) What is meant by a 'non-renewable' resource?
          (1 mark)

3 It is now possible to 'clone' cats. This means that if your pet cat dies you can create an identical copy. It is a difficult and expensive process. The diagram below shows how it can be done.

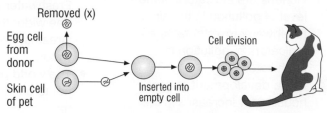

(a) (i) Explain, as fully as you can, why the 'cloned' cat is genetically identical to the original pet. (4 marks)
   (ii) What name is given to this type of cloning?
          (1 mark)

(b) Many people are against cloning animals.
   Give two reasons why people might be against cloning animals.   (2 marks)

(c) It is more difficult to clone animals than it is to clone plants.
   State two different ways that plants can be cloned. For each way, explain how it is carried out.   (4 marks)

4 Most scientists now accept Darwin's theory of evolution.

(a) (i) How does Charles Darwin's theory of evolution explain the development, over many generations, of giraffes with increasingly long necks? (4 marks)
   (ii) Lamarck had previously put forward a completely different explanation for evolution. How would Lamarck have explained the development of giraffes with increasingly long necks?   (2 marks)

(b) (i) How long ago do we believe that life began on Earth?   (1 mark)
   (ii) What are the two different theories about how life first began on the Earth?   (4 marks)

(c) How do we know when the organisms, represented by fossils, were present on the Earth?   (2 marks)

(d) There are a number of reasons why species have become extinct.
   State two reasons and, for each, explain why this could have resulted in the species becoming extinct.
          (2 marks)

Test & Assessment Interactive quizzes, answers and hints online!

The answer is worth 4 marks out of the 5 available. The responses worth a mark are underlined in red.

We can improve the answer in several ways:

Without the 'greenhouse effect', the Earth would be much cooler than it is. Explain, as fully as you can, how the higher concentrations of methane and carbon dioxide are affecting the Earth and causing it to become warmer.

(5 marks)

The carbon dioxide and methane are trapped by the Earth's atmosphere. The gases have a warming effect. They trap the Sun's rays within the atmosphere preventing them from escaping back into space. There are now more extreme weather conditions than there used to be. The polar ice caps might melt and cause the seas to rise and an influx of water and flooding.

The first two sentences of the answer were not worth a mark and therefore the student wasted time here.

A mark would be gained by stating that some species could **become extinct** if their habitat disappears, e.g. polar bears if the Arctic ice cap melts.

A mention of the **changing distribution** of species would gain a mark, e.g. certain plants or animals found in Southern Europe could, in future, be found in the UK.

A fifth mark could have been gained by stating that **more** of the Sun's rays would not be reflected back into space (some still will be).

The answer is worth 3 marks out of the 4 available. The responses worth a mark are underlined in red.

We can improve the answer in several ways:

Sulfur dioxide is sometimes released when fossil fuels are burned. Describe, in as much detail as you can, how this might affect the environment.

(4 marks)

Sulfur dioxide pollutes the environment. Factories burn fossil fuels which give off sulfur dioxide, carbon dioxide and nitrogen oxide. It dissolves in water in the clouds. This can get into lakes and rivers and makes the water more acidic, it also kills the animals and plants. Also it heats up the planet so the extra heat cannot get back into space.

A mention of **acid rain** or the wind blowing the clouds to other areas would have been worth the fourth mark.

The 'greenhouse effect' (carbon dioxide) is not relevant to the question. Sometimes by giving too much information, especially when it is wrong, will result in you losing some of the marks.

There was no need to mention carbon and nitrogen oxides, as they are not relevant to the question and made worse by giving incorrect information, e.g. carbon dioxide does not get into lakes.

Checklist

This spider diagram shows the topics in the first half of the unit. You can copy it out and add your notes and questions around it, or cross off each section when you feel confident you know it for your exams.

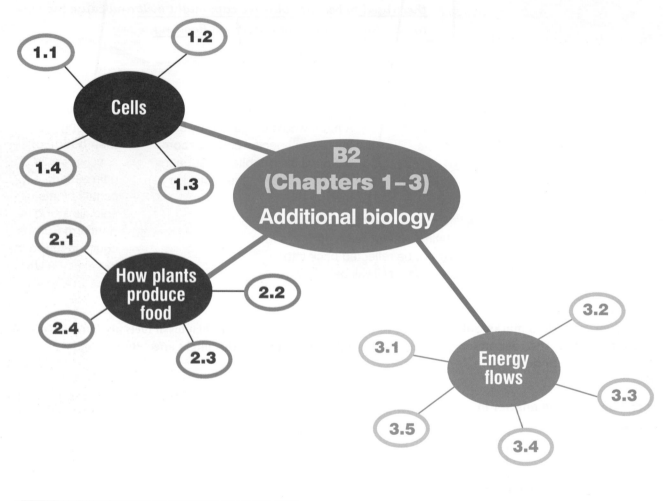

Tick when you:

reviewed it after your lesson	✓	☐	☐
revised once – some questions right	✓	✓	☐
revised twice – all questions right	✓	✓	✓

Move on to another topic when you have all three ticks.

Chapter 1 Cells

1.1	Animal and plant cells	☐	☐	☐
1.2	Specialised cells	☐	☐	☐
1.3	How do substances get in and out of cells?	☐	☐	☐
1.4	Osmosis	☐	☐	☐

Chapter 2 How plants produce food

2.1	Photosynthesis	☐	☐	☐
2.2	Limiting factors	☐	☐	☐
2.3	How plants use glucose	☐	☐	☐
2.4	Why do plants need minerals?	☐	☐	☐

Chapter 3 Energy flows

3.1	Pyramids of biomass	☐	☐	☐
3.2	Energy losses	☐	☐	☐
3.3	Energy in food production	☐	☐	☐
3.4	Decay	☐	☐	☐
3.5	The carbon cycle	☐	☐	☐

What are you expected to know?

Chapter 1 Cells (See students' book pages 138–147)

- Animal and plant cells contain a nucleus, cell membrane, cytoplasm, mitochondria and ribosomes.

- The functions (jobs) of these different organelles.

- Plant cells also have a cell wall, chloroplasts and a permanent vacuole.

- The functions of these organelles in plants.

- Some cells are specialised to carry out their function.

- Materials are transported across cell membranes by diffusion and osmosis.

- How the processes of diffusion and osmosis are similar and what the differences are between them.

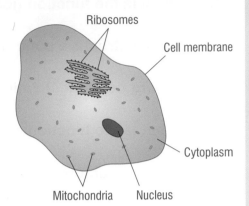

Chapter 2 How plants produce food

See students' book pages 150–159

- Plants produce carbohydrates through photosynthesis and some of these are stored.

- The factors that limit the rate of photosynthesis are the:
  - level of carbon dioxide
  - temperature
  - light intensity.

- Plants need nitrate and magnesium in order to grow properly.

- If a plant is deficient in these nutrients then the plant will show symptoms.

Chapter 3 Energy flows (See students' book pages 162–173)

- A pyramid of biomass tells you the mass of the different organisms in a food chain (and it can give you a better picture of a food chain than a pyramid of numbers).

- Energy is transferred in a food chain from one organism to the next.

- Energy is lost in food chains.

- The shorter a food chain is the more efficient food production will be.

- We can artificially improve the efficiency of a food chain when we are producing food, but this can be controversial.

- Materials are recycled by microorganisms.

- A stable community recycles all of the nutrients.

- The carbon cycle.

(1) What is the function (job) of the nucleus?

(2) Which structure controls the movement into and out of a cell?

(3) Where do most chemical reactions take place in a cell?

(4) What is the function of ribosomes?

(5) Which structures are necessary for photosynthesis to take place?

(6) Which structures are necessary for aerobic respiration to take place?

(7) Why are some plant and animal cells specialised?

(8) How can diffusion be defined?

(9) What substance(s) might diffuse into or out of cells?

(10) How is osmosis different to diffusion?

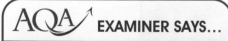

students' book
page 138

B2 1.1 Animal and plant cells

KEY POINTS

1 Animal and plant cells have structures that enable them to do their jobs.
2 Plant cells have some structures which animal cells don't have.

AQA EXAMINER SAYS...

Remember that the cell membrane can control the movement of materials into and out of the cell. The cell wall, in plants, does not do this. It is there for support.

Animal and plant cells have some structures in common; they have:

- a nucleus to control the cell's activities
- cytoplasm where many chemical reactions take place
- a cell membrane that controls the movement of materials
- mitochondria where energy is released during aerobic respiration
- ribosomes where proteins are made (synthesised).

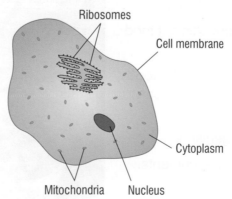

A simple **animal cell** like this shows the features that are common to all living cells

Plant cells also have:

- a rigid cell wall for support
- chloroplasts that contain chlorophyll for photosynthesis
- a permanent vacuole containing cell sap.

Key words: nucleus, cytoplasm, cell membrane, mitochondria, ribosomes, cell wall, chloroplasts, vacuole

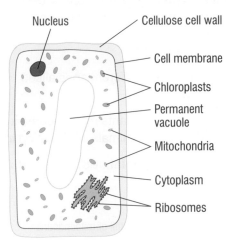

Nucleus

Cellulose cell wall

Cell membrane

Chloroplasts

Permanent vacuole

Mitochondria

Cytoplasm

Ribosomes

A **plant cell** has many features in common with an animal cell, but others which are unique to plants

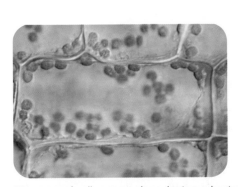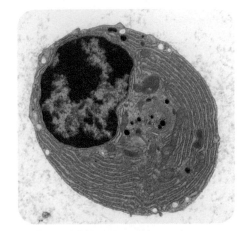

Diagrams of cells are much easier to understand than the real thing seen under a microscope. These pictures show a magnified plant cell and animal cell.

CHECK YOURSELF

1 Name three structures common to both plant and animal cells.

2 Where does aerobic respiration take place?

3 Name two materials that move across the cell membrane.

B2 1.2 Specialised cells

KEY POINTS

1 As organisms develop, some of their cells become specialised to carry out particular jobs. This is called 'differentiation'.
2 Differentiation happens much earlier in the development of animals than it does in plants.

When an egg is fertilised it begins to grow and develop.

At first there is a growing ball of cells. Then as the organism gets bigger some of the cells change and become specialised.

There are many different specialised cells, e.g.

- Some cells in plants may become xylem or root hair cells.
- Some cells in animals will develop into nerve or sperm cells.

Key words: specialised, differentiation

EXAM HINTS

In an exam it is not only important to remember that there are specialised cells. It is just as important to remember how their structure makes them suitable to carry out the job that they do.

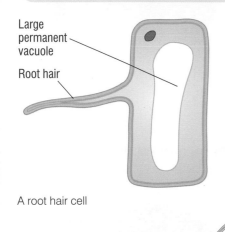

Large permanent vacuole

Root hair

A root hair cell

Middle section – full of mitochondria

Nucleus

Tail

A sperm cell

CHECK YOURSELF

1 Can you give two examples of specialised animal cells not mentioned here?
2 How is the structure of a nerve cell related to its function?
3 What do we call the process that results in some cells becoming specialised?

B2 1.3 How do substances get in and out of cells?

KEY POINT

Diffusion is the result of random movement. It does not require any energy from the cell.

Molecules move randomly because of the energy they have.

Diffusion is the random movement of molecules from an area of *high concentration* to an area of *lower concentration.*

The larger the difference in concentration, the faster the rate of diffusion. Examples are:

- the diffusion of oxygen into the cells of the body from the blood stream as the cells are respiring (and using up oxygen)
- the diffusion of carbon dioxide into actively photosynthesising plant cells
- the diffusion of simple sugars and amino acids from the gut through cell membranes.

Key words: random, diffusion, concentration

GET IT RIGHT!

Remember that diffusion does not require energy from the cells. It is simply the random movement of particles. The greater the difference in concentrations, the faster diffusion takes place.

CHECK YOURSELF

1 Why do particles move randomly?
2 Can you think of another example of diffusion in living cells?
3 Why is diffusion important?

B2 1.4 Osmosis

KEY POINT

Osmosis is a special case of diffusion involving a partially (or semi-) permeable membrane.

EXAMINER SAYS...

Remember osmosis is just a special case of diffusion. It only involves the movement of water and it takes place across a partially permeable membrane.

Osmosis is the movement of water.

Just like diffusion, the movement of molecules is random and requires no energy from the cell.

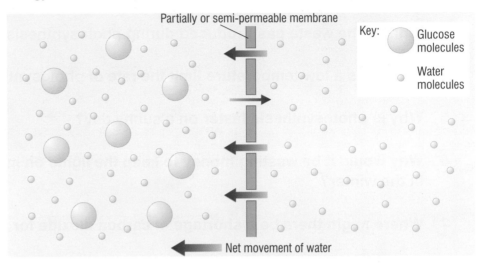

Partially or semi-permeable membrane

Key:
Glucose molecules
Water molecules

Net movement of water

Osmosis

CHECK YOURSELF

1 How is osmosis like diffusion?

2 What do you think 'partially permeable' means?

3 Why do cells need water?

Osmosis is the diffusion of water across a partially permeable membrane from a dilute solution to a more concentrated solution. No solute molecules can move across the membrane. The cell membrane is partially permeable.

Water is needed to support cells and because chemical reactions take place in solution.

Key words: osmosis, partially permeable, solute

B2 1 End of chapter questions

1 **What is the function of ribosomes?**

2 **Suggest how a sperm cell is adapted to its function.**

3 **Why does diffusion require no energy from the cell?**

4 **Why would it be incorrect to say that in osmosis water only passes from the weaker to the stronger solution?**

5 **What substance, necessary for photosynthesis, do chloroplasts contain?**

6 **What is the function of the cell wall in plants?**

7 **Why does oxygen in the blood diffuse into respiring cells?**

8 **Which part of a cell's structure is 'partially permeable'?**

1. Where does the energy for photosynthesis come from?

2. Which carbohydrate is produced during photosynthesis?

3. What is the waste gas produced during photosynthesis?

4. Why does a low temperature limit the rate of photosynthesis?

5. Why is photosynthesis faster on a sunny day?

6. Why would it be wasting money to keep the lights on in a very cold greenhouse in the winter?

7. Where might there be a shortage of carbon dioxide for photosynthesis?

8. Why do plants need nitrate?

9. What would a plant that is short of nitrate look like?

10. Why do plants short of magnesium have yellow leaves?

---

students' book page 150

B2 2.1   Photosynthesis

KEY POINTS

1 Photosynthesis can only be carried out by green plants.
2 Chlorophyll traps the Sun's energy.

GET IT RIGHT!

Remember that chlorophyll *traps* the Sun's energy, it does not produce any energy of its own.

The equation for photosynthesis is:

carbon dioxide + water **(+ light energy)** → glucose + oxygen

The carbon dioxide is taken in by the leaves, and the water by the roots.

The chlorophyll traps the energy needed for photosynthesis.

In photosynthesis the sugar glucose (a carbohydrate) is made. Oxygen is given off as a waste gas.

These leaves came from a plant which had been kept in the light for several hours. Only the green parts of the leaf made their own starch which turns the iodine solution blue-black.

The structure of a leaf

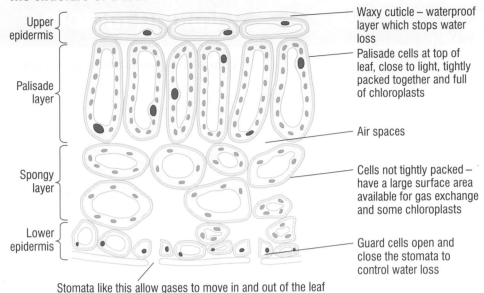

Upper epidermis

Palisade layer

Spongy layer

Lower epidermis

Waxy cuticle – waterproof layer which stops water loss

Palisade cells at top of leaf, close to light, tightly packed together and full of chloroplasts

Air spaces

Cells not tightly packed – have a large surface area available for gas exchange and some chloroplasts

Guard cells open and close the stomata to control water loss

Stomata like this allow gases to move in and out of the leaf

Key words: photosynthesis, energy

EXAM HINT

Learn the equation for photosynthesis . . . and make sure you can explain the results of experiments on photosynthesis.

CHECK YOURSELF

1 Where does the energy for photosynthesis come from?

2 What are the two substances necessary to make the glucose?

3 Why is the waste product of photosynthesis so important?

---

students' book page 152

B2 2.2   # Limiting factors

KEY POINT

If certain things are in short supply, they will slow down the rate of photosynthesis. Plant growers need to know this, otherwise they could waste money.

A lack of light would slow down the rate of photosynthesis as light provides the energy for the process. Chlorophyll traps the light. Even on sunny days, light may be limited on the floor of a wood or rain forest.

If it is cold, then enzymes do not work effectively and this will slow down the rate.

If there is too little carbon dioxide, then the rate will slow down. Carbon dioxide may be limited in an enclosed space, e.g. in a greenhouse on a sunny day or in a rapidly photosynthesising rain forest.

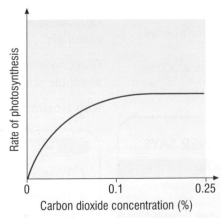

This graph shows the effect of increasing carbon dioxide levels on the rate of photosynthesis at a particular light level and temperature. Eventually one of the other factors becomes limiting.

Look at the graphs below showing the other two limiting factors:

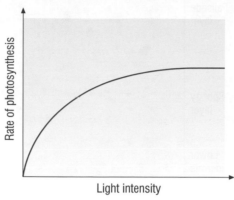

The effect of light intensity on the rate of photosynthesis

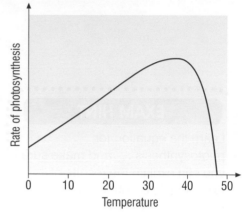

The rate of photosynthesis increases steadily with a rise in temperature up to a certain point. After this the enzymes are destroyed and the reaction stops completely.

For anyone growing plants, there is no point increasing one of these factors if the photosynthesis is still limited by another one. They would just be wasting money.

Key words: limiting, rate

CHECK YOURSELF

1 Why does photosynthesis slow down on a cold day?

2 Why would there be little point heating a greenhouse on a summer's day?

3 Other than in an enclosed space (e.g. a greenhouse) where might photosynthesis be limited by the amount of CO_2?

How plants use glucose

KEY POINTS

1 Plants use glucose for energy (respiration).
2 Most plants can store glucose as starch.

The product of photosynthesis is glucose. Glucose is used for respiration.

Glucose is also combined with other nutrients by the plant to produce new materials.

Glucose is stored, by some plants, as insoluble starch. It is stored as an insoluble substance so that it has no effect on osmosis.

Key words: glucose, starch, insoluble

CHECK YOURSELF

1 Why is glucose stored as an insoluble compound?

2 Glucose is used for respiration. What else is it used for?

3 What is meant by 'respiration'?

B2 2.4 Why do plants need minerals?

KEY POINTS

1 Plants produce sugars through photosynthesis.
2 However, just like animals, they need other substances to grow properly.

Plant roots take up mineral salts for healthy growth.

Nitrates are taken from the soil for producing amino acids. These are used to make proteins for growth. A plant that does not take up enough nitrate (is nitrate deficient) will have stunted growth.

Plants also take up magnesium that is essential to produce chlorophyll. If the plant is deficient in chlorophyll it will have yellow leaves.

Key words: nitrate, magnesium, deficient

GET IT RIGHT!

Many students in an exam forget the symptoms of nitrate and magnesium deficiency, or get them the wrong way round. Try to work out a way of memorising them, e.g. nite/stunt and yellow/mag!

The plants on the left of this picture have been grown in a mixture containing all the minerals they need. The experimental plants on the right have been grown without nitrates. The difference in their rate of growth is clear to see.

CHECK YOURSELF

1 How do plants take minerals from the soil?

2 Why do plants need magnesium?

3 What does 'deficiency' mean?

B2 2 End of chapter questions

1 **How is light energy trapped by a plant?**

2 **You are growing some cuttings in the greenhouse. It is a cold, sunny winter's day. What factor will be limiting the growth of the cuttings?**

3 **Why is glucose stored as an insoluble compound?**

4 **You grow two plants of the same species in the same conditions but one is deficient in nitrate. How will its growth compare with the growth of the healthy plant?**

5 **Which carbohydrate is produced directly as a result of photosynthesis?**

6 **Other than in woodland, where might light be a limiting factor for some plants?**

7 **Why do plants not grow properly if there is not enough nitrate in the soil?**

8 **How do most plants store glucose?**

① **What is a pyramid of biomass?**

② **Why does a pyramid of biomass often give a more accurate picture of what is happening than a pyramid of numbers?**

③ **Why is energy lost between the different stages of a food chain?**

④ **Why is more energy lost when birds are part of the food chain?**

⑤ **Why do we say, in terms of food production, that meat is 'expensive' to produce?**

⑥ **How can you limit the energy losses in a food chain?**

⑦ **What other word do we use for the breakdown of dead plants and animals?**

⑧ **Which organisms break down dead and waste material?**

⑨ **Which process results in carbon dioxide being taken out of the atmosphere?**

⑩ **What do we mean by a 'stable' community in terms of the recycling of nutrients?**

---

students' book page 162

B2 3.1   Pyramids of biomass

KEY POINTS

1 A pyramid of biomass represents the mass of all the organisms at each stage in a food chain.
2 A pyramid of biomass is likely to give a more accurate picture than a pyramid of numbers.

Biomass is the mass of living material in plants and animals.

A pyramid of biomass represents the mass of the organisms at each stage in a food chain. It may be more accurate than a pyramid of numbers. For example, one bush may have many insects feeding on it but the mass of the bush is far greater than the mass of the insects.

Organism	Number	Biomass – dry mass in g
Oak tree	1	500 000
Aphids	10 000	1000
Ladybirds	200	50

Pyramid of numbers   Pyramid of biomass

This food chain cannot be accurately represented using a pyramid of numbers. Using biomass shows us the amount of biological material involved at each level in a way that simple numbers cannot do.

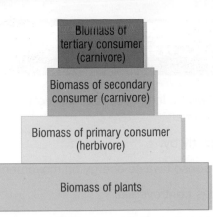

Any food chain can be turned into a pyramid of biomass like this

You can draw pyramids of biomass to scale to give an even more accurate picture.

Key words: pyramid, biomass

AQA EXAMINER SAYS...

Make sure you can explain why a pyramid of biomass is likely to give a more accurate picture of a food chain than a pyramid of numbers.

CHECK YOURSELF

1 What does the word 'biomass' mean?

2 Can you think of another example where a pyramid of biomass will give a more accurate picture of a food chain than a pyramid of numbers?

3 What does drawing a pyramid 'to scale' mean?

---

students' book page 164

B2 3.2 Energy losses

KEY POINTS

1 For a whole range of reasons, there is energy loss between each stage of a food chain.

2 This means that not all of the energy taken in by an organism results in the growth of that organism.

Not all of the food eaten can be digested, so energy is lost in faeces (waste materials).

Some of the energy is used for respiration, which releases energy for living processes. This includes movement, so the more something moves the more energy it uses and the less is available for growth.

In animals that need to keep a constant temperature, energy from the previous stage of the food chain is used simply to keep the animal at the correct temperature (e.g. 37°C in humans).

AQA EXAMINER SAYS...

Make sure that you can interpret Sankey diagrams. You may well get one in the exam.

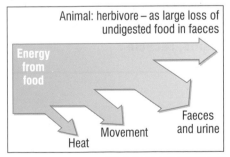

Animal: herbivore – as large loss of undigested food in faeces

Energy from food

Heat Movement Faeces and urine

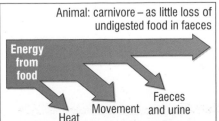

Animal: carnivore – as little loss of undigested food in faeces

Energy from food

Heat Movement Faeces and urine

GET IT RIGHT!

Energy is never really 'lost'. What we mean here is that all of the energy in one stage of the food chain does not result in the growth of organisms in the next stage.

Sankey diagrams show how energy is transferred in a system. We can use them to look at the energy which goes in to and out of an animal and predict whether it eats plants or is a carnivore.

CHECK YOURSELF

1 What is meant by 'energy loss' in a food chain?

2 Why does maintaining a constant body temperature use a lot of energy from food?

3 Why does a running horse need more energy than one eating grass in a field?

B2 3.3 Energy in food production

EXAMINER SAYS...

There are issues in this topic. One is that if we all ate plants it might mean more food would be available across the world. A second is that if we keep animals so that they produce meat efficiently for us this could be cruel to the animals. Make sure that you have an opinion, but also that you can give both sides of the argument.

The shorter the food chain, the less energy will be lost. It is therefore more efficient for us to eat plants than it is to eat animals.

We can artificially produce meat more efficiently by:

- Preventing the animal from moving so it doesn't waste energy on movement.

This is seen as cruelty by many people and is controversial.

- Keeping the animal at a warmer temperature so it doesn't use as much energy from food to keep itself at a constant temperature.

Key words: efficient

CHECK YOURSELF

1 What do we mean by 'more efficient' food production?

2 If an animal has a constant internal temperature, why might it use more energy when grazing in a field in winter?

3 Why are some methods of producing meat more efficiently said to be cruel?

B2 3.4 Decay

GET IT RIGHT!

If microorganisms are feeding from waste and dead material, then this is how they get their energy. The process of releasing this energy is, therefore, respiration so carbon dioxide is given off.

Detritus feeders (such as worms) may start the process of decay by eating dead animals or plants and producing waste materials. Decay organisms then break down the waste and dead plants and animals.

Decay organisms are microorganisms (bacteria and fungi). Decay is faster if it is warm and wet.

All of the materials from the waste and dead organisms are recycled.

Key words: detritus, decay, microorganisms, recycle

CHECK YOURSELF

1 What is a detritus feeder?

2 Why is decay faster when it is warm?

3 Suggest another word or phrase for 'decay'.

4 Give one example of a microorganism.

B2 3.5   The carbon cycle

EXAMINER SAYS...

If you can remember that the carbon cycle involves both photosynthesis and respiration, then you will be awarded most of the marks in an exam question.

CHECK YOURSELF

1 Which process returns CO_2 to the atmosphere?

2 Which organisms break down waste products?

3 What does 'recycling carbon' mean?

Photosynthesis removes CO_2 from the atmosphere.

Green plants as well as animals respire. This returns CO_2 to the atmosphere.

Animals eat green plants and build the carbon into their bodies. When plants or animals die (or produce waste) microorganisms release the CO_2 back into the atmosphere through respiration.

A stable community recycles all of the nutrients it takes up.

Key words: photosynthesis, respiration, recycles, stable

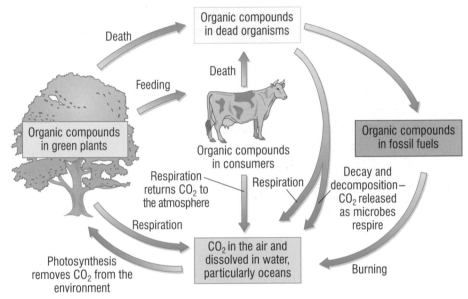

The carbon cycle in nature

B2 3   End of chapter questions

1 **What process is necessary to provide the energy needed to keep humans at a constant temperature?**

2 **Why is very little meat eaten by poorer people in developing countries?**

3 **Why do microorganisms recycle nutrients more quickly when it is warm?**

4 **What involvement do living plants have in the carbon cycle?**

5 **Why does a pyramid of biomass usually give a more accurate picture of what is happening in a community than a pyramid of numbers?**

6 **What happens to the food that animals do not digest?**

7 **Which group of organisms often start the decay process by eating dead plants and animals?**

8 **At what time of the day do plants respire?**

1 The diagram shows a nerve cell.

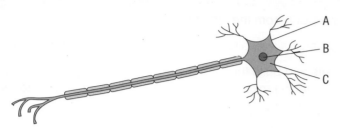

(a) Name parts A, B and C.   (3 marks)

(b) The nerve cell carries impulses (electrical messages) around the body. Write down two ways the structure of the cell helps it carry out this function (job).
                                          (2 marks)

(c) (i) Name two structures a plant cell has and an animal cell does not have.
    (ii) For each structure you have named in part (i), describe one of its functions.   (4 marks)

2 You are walking in a wood. You see large numbers of caterpillars feeding on the leaves of the trees and some small birds (blue tits) eating the caterpillars. Suddenly a fairly large bird (a sparrowhawk) catches one of the blue tits and starts eating it.

(a) The diagram shows a pyramid of biomass for the organisms you have seen.
    Write the names of the organisms you have seen in the woods in the correct places next to the pyramid of biomass.   (3 marks)

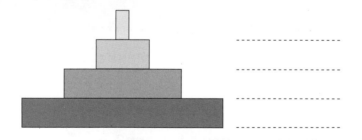

(b) Why is so much energy lost between each of the stages in this woodland food chain?   (3 marks)

(c) Producing meat more efficiently means that more of the food that the animals eat is converted into meat and not wasted. How can we produce meat more efficiently?   (2 marks)

3 The photograph shows a very small area of the Amazon jungle in Ecuador, South America. The growth of plants is very fast in the jungle.

(a) The plants in the jungle photosynthesise and produce glucose in their leaves.
    (i)   What do the plants use the glucose for? (1 mark)
          When does this process take place?   (1 mark)
    (ii) The plants will also be taking up nitrates from the soil.
          What do they use these nitrates for?   (2 marks)
    (iii) If there is not enough nitrate then the plant will have a deficiency. What will the plant look like?
                                          (1 mark)

(b) In an area where plants are growing so quickly, they may not be able to take up enough magnesium. Explain the results of magnesium deficiency in a growing plant.   (2 marks)

(c) When the plants die they are recycled.
    What conditions increase the speed at which recycling takes place?   (2 marks)

(d) Except where humans are present, the Amazon jungle is said to be a 'stable' community. What does this mean?   (1 mark)

4 Cells need to take up materials that they need and get rid of waste materials from the reactions that have taken place.

(a) (i) Diffusion is one way that materials pass into and out of cells.
          What is meant by diffusion?   (3 marks)
    (ii) In a plant cell which is photosynthesising which gas, overall, will be:
          A diffusing into the plant cells
          B diffusing out of the plant cells?   (2 marks)

(b) Why do the cells not need to provide energy for the process of diffusion?   (2 marks)

(c) Osmosis is a 'special' case of diffusion.
    (i)   How is osmosis similar to diffusion?   (2 marks)
    (ii) How is osmosis different to diffusion?   (2 marks)

 Test & Assessment Interactive quizzes, answers and hints online!

The answer is worth 3 marks out of a possible 4. The responses worth a mark are underlined in red.

We can improve the answer in several ways:

It would improve the answer to state that the light provides the energy for the process.

The comment is too vague to be worth a mark. A fuller explanation is required, e.g. the greenhouse will be warm as it is sunny, the enzymes controlling photosynthesis will, therefore, be working quickly.

The diagram shows a greenhouse on a sunny summer's day.

Sun

Which one of these factors will limit the rate of photosynthesis in the greenhouse?

- **the level of carbon dioxide**
- **the level of light**
- **the temperature**

Explain your answer. *(4 marks)*

The carbon dioxide level.
The plants will photosynthesise quickly as there should be enough light as it is a sunny day. The temperature will also be warm enough. If all of these things are going quickly enough then the plants will use up the carbon dioxide quickly so there won't be enough.

The answer is worth 2 marks out of the 4 marks available. The responses worth a mark are underlined in red.

We can improve the answer in several ways:

The 'little animals' mentioned would gain the mark if 'bacteria or fungi', or the general term 'microorganisms', had been given.

'Detritus feeders' feed on the leaves (digest them) and start the process of breakdown, e.g. earthworms.

In the Autumn many plants lose their leaves for the winter. There are carbon compounds in the leaves. How is this carbon recycled so it can be used again by plants? *(4 marks)*

When the leaves fall little animals come and eat them and through respiration energy and carbon dioxide are released. Then other little animals feed on the waste and the whole process is repeated.

The process of breakdown is also known as 'decomposition' or 'decay'.

Additional biology (Chapters 4–6)

Checklist

This spider diagram shows the topics in the second half of the unit. You can copy it out and add your notes and questions around it, or cross off each section when you feel confident you know it for your exams.

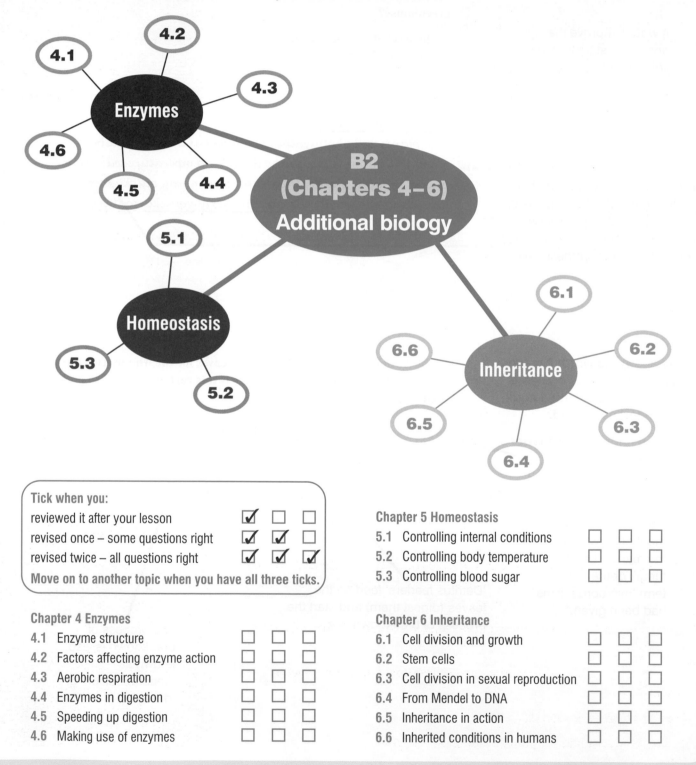

Tick when you:

reviewed it after your lesson	☑	☐	☐
revised once – some questions right	☑	☑	☐
revised twice – all questions right	☑	☑	☑

Move on to another topic when you have all three ticks.

Chapter 4 Enzymes

4.1	Enzyme structure	☐	☐	☐
4.2	Factors affecting enzyme action	☐	☐	☐
4.3	Aerobic respiration	☐	☐	☐
4.4	Enzymes in digestion	☐	☐	☐
4.5	Speeding up digestion	☐	☐	☐
4.6	Making use of enzymes	☐	☐	☐

Chapter 5 Homeostasis

5.1	Controlling internal conditions	☐	☐	☐
5.2	Controlling body temperature	☐	☐	☐
5.3	Controlling blood sugar	☐	☐	☐

Chapter 6 Inheritance

6.1	Cell division and growth	☐	☐	☐
6.2	Stem cells	☐	☐	☐
6.3	Cell division in sexual reproduction	☐	☐	☐
6.4	From Mendel to DNA	☐	☐	☐
6.5	Inheritance in action	☐	☐	☐
6.6	Inherited conditions in humans	☐	☐	☐

What are you expected to know?

Chapter 4 Enzymes (See students' book pages 176–189)

- How enzymes work in terms of their structure.
- That different enzymes work best under different conditions of pH.
- How aerobic respiration releases energy.
- The role of enzymes in the digestion of carbohydrate, protein and fat (lipids).
- The role of the stomach, pancreas, small intestine and liver in the digestion of food.
- How enzymes are used in the home.
- How enzymes are used in industry.

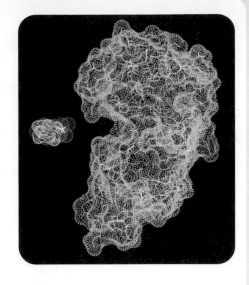

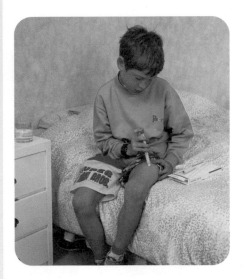

Chapter 5 Homeostasis (See students' book pages 192–199)

- How we get rid of the waste products carbon dioxide and urea.
- How we keep our internal conditions constant including:
  - water content
  - ion content
  - temperature
  - blood sugar level.

Chapter 6 Inheritance (See students' book pages 202–215)

- How characteristics are passed on from one generation to the next (inheritance).
- What happens in mitosis.
- What happens during meiosis. [Detail and comparisons with mitosis are Higher Tier only]
- The issues surrounding the use of stem cells (from embryos and adult bone marrow).
- That a gene is a short length of DNA controlling one characteristic, that pairs of genes controlling the same characteristic are known as 'alleles' and that chromosomes are made up of many genes.
- That some disorders, e.g. Huntingdon's disease and cystic fibrosis, are inherited.
- That embryos can be screened for these disorders and other genetic disorders.
- How to interpret genetic diagrams.
- How to draw genetic diagrams. [Higher Tier only]

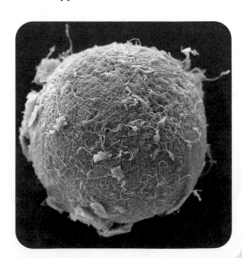

1. What do we mean by the term 'biological catalyst'?

2. State two processes that are controlled by enzymes.

3. Why is the shape of an enzyme important?

4. Where does aerobic respiration take place?

5. In living cells what are amino acids built up into?

6. What is the general name for an organ which produces an enzyme?

7. What type of enzyme breaks down fats?

8. Where is bile produced?

9. Where is hydrochloric acid produced?

10. Why is fructose such a useful sugar?

---

students' book page 176

B2 4.1   Enzyme structure

KEY POINTS

1. The structure of an enzyme allows certain molecules to fit.
2. If this structure is changed (the enzyme is denatured) then the enzyme cannot work as a catalyst.

GET IT RIGHT!

Many students in the exam talk of enzymes being 'killed' by high temperatures. This answer is worth *no marks* – they are destroyed or denatured.

Enzymes are biological catalysts – they speed up reactions.

Enzymes are large proteins and each has a particular shape. This shape has an area where other molecules can fit in. This area is called the 'active site'.

Too high a temperature will change the enzymes shape, and it will no longer work. We say it has been destroyed or denatured.

Enzymes can catalyse the build up of small molecules into large molecules or the break down of large molecules into small molecules.

Enzymes lower the amount of energy necessary for a reaction to take place – the 'activation' energy.

Key words: catalyst, active site, denatured, activation energy

CHECK YOURSELF

1. What type of molecules are enzymes?
2. What is the 'active site'?
3. What does 'catalysis' mean?

B2 4.2 Factors affecting enzyme action

EXAMINER SAYS...

If you get a question about rate of reaction, remember to talk about collisions. How hard and how often molecules collide determines the rate of reaction. Don't be vague – mention 'collisions'!

Reactions take place faster when it is warmer. At the higher temperature the molecules move around more quickly so collide with each other more often and with more energy.

However, if the temperature gets too hot the enzyme stops working. That's because the active site changes shape and the enzyme becomes denatured.

Enzymes work best in certain acidic or alkaline conditions (pH). If the pH is too acidic or alkaline for the enzyme, then the active site could change shape. The enzyme would stop working.

Key words: collide, active site

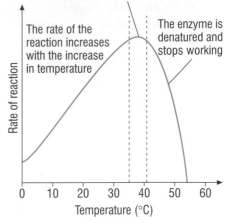

Like most chemical reactions, the rate of an enzyme controlled reaction increases as the temperature rises – but only until the point where the complex protein structure of the enzyme breaks down

CHECK YOURSELF

1 Why does an increase in temperature increase the rate of reaction?

2 Why do enzymes stop working if the temperature is too hot?

3 Why does the pH have to be 'right' for a reaction to go well?

B2 4.3 Aerobic respiration

EXAM HINT

Remember that animals and plants both respire, it is the same process. Plants do not just photosynthesise.

The equation for respiration is:

glucose + oxygen → carbon dioxide + water [+energy]

The process mostly takes place in the mitochondria.

The energy released is used to:

- build larger molecules from smaller ones
- enable muscle contraction in animals
- maintain a constant body temperature in mammals and birds
- build sugars, nitrates and other nutrients in plants into amino acids and then proteins.

Key words: mitochondria, energy

CHECK YOURSELF

1 Where does aerobic respiration take place?

2 What is needed for muscles to contract?

3 What are amino acids built up into?

B2 4.4 Enzymes in digestion

KEY POINTS

1 Without enzymes, digestion would be too slow.
2 There are specific conditions in different parts of the gut that help enzymes to work effectively.

Digestion involves the breakdown of large, insoluble molecules into smaller soluble molecules.

- Amylase (a carbohydrase) is produced by the salivary glands, the pancreas and the small intestine. Amylase catalyses the digestion of starch into sugars in the mouth and small intestine.
- Protease is produced by the stomach, the pancreas and the small intestine. Protease catalyses the breakdown of proteins into amino acids in the stomach and small intestine.
- Lipase is produced by the pancreas and small intestine. Lipase catalyses the breakdown of lipids (fats and oils) to fatty acids and glycerol.

Key words: amylase, protease, lipase, lipid, amino acid, fatty acid, glycerol

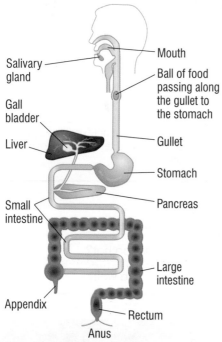

Salivary gland
Mouth
Gall bladder
Ball of food passing along the gullet to the stomach
Liver
Gullet
Stomach
Small intestine
Pancreas
Large intestine
Appendix
Rectum
Anus

The human digestive system

CHECK YOURSELF

1 What are lipids?
2 Which enzymes does the pancreas produce?
3 What are the products of lipase digestion?

B2 4.5 Speeding up digestion

GET IT RIGHT!

Large numbers of students do not remember that bile is made by the liver and stored in the gall bladder. Try to get it right!

- Protease enzymes in the stomach work best in acid conditions. Glands in the stomach wall produce hydrochloric acid to create very acidic conditions.
- Amylase and lipase work in the small intestine. They work best when the conditions are slightly alkaline.
- The liver produces bile that is stored in the gall bladder. Bile is squirted into the small intestine and neutralises the stomach acid. It makes the conditions slightly alkaline.

Key words: acid, alkali, bile, liver, gall bladder

CHECK YOURSELF

1 What conditions are best in the small intestine?
2 What is the function of bile?
3 What type of acid is produced in the stomach?

Making use of enzymes

KEY POINTS

1 Enzymes can be used in products in the home and in industry.
2 Microorganisms produce enzymes that we can use.

Learning to eat solid food isn't easy. Having some of it pre-digested by protease enzymes can make it easier to get the goodness you need to grow!

Biological washing powders contain enzymes that digest food stains. They work at lower temperatures than ordinary washing powders so can save us money.

We also use:

- Protease enzymes to pre-digest proteins in some baby foods.
- Isomerases to convert glucose into fructose. Fructose is much sweeter, so less is needed in foods. The foods, therefore, are not so fattening.
- Carbohydrases to convert starch into sugar syrup for use in foods.

Key words: isomerase, fructose, pre-digest

AQA EXAMINER SAYS...

Try to remember at least two examples of these uses of enzymes. Most questions will not ask for more than two.

CHECK YOURSELF

1 Why is fructose used in slimming foods?

2 Suggest why baby food might need some of the protein content 'pre-digested'.

3 Why would you not use a temperature of above about 45°C if you are using a biological washing powder?

B2 4   End of chapter questions

1 What do we mean by the 'activation energy'?

2 What is the 'active site' of an enzyme?

3 State two uses, in animals, of the energy released by respiration.

4 What digestive process takes place in the stomach?

5 What term do we use when the active site of an enzyme changes shape?

6 What are the two waste products of aerobic respiration?

7 Where is protease produced in the digestive system?

8 Where is bile made and where is it stored?

1. What are the waste products of respiration?

2. What happens to the amino acids we cannot use?

3. Where is urea produced?

4. Why is the ion content of the body important?

5. Which two areas of the body detect changes in temperature?

6. Why does sweating more cool us down? [Higher Tier only]

7. Where is sweat produced? [Higher Tier only]

8. What effect does 'shivering' have on the body? [Higher Tier only]

9. Which hormone helps to control the blood sugar level?

10. Where is this hormone produced?

---

students' book
page 192

B2 5.1 Controlling internal conditions

KEY POINTS

1 We must remove the waste products produced through chemical reactions from the body.

2 There are other factors we must keep within certain limits, e.g. water and ion content of the cells.

The processes in your body that help to maintain a constant internal environment are known as homeostasis.

Just think of the different temperatures we experience:

Your body also has to cope with varying rates of respiration needed in different activities:

Whatever you choose to do in life, the conditions inside your body will stay more-or-less the same

We can investigate the effect of exercise on our breathing using the equipment below:

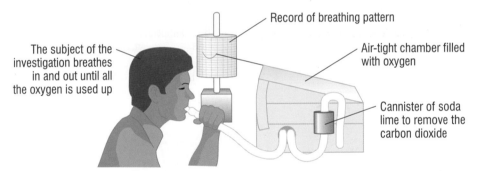

The subject of the investigation breathes in and out until all the oxygen is used up

Record of breathing pattern

Air-tight chamber filled with oxygen

Cannister of soda lime to remove the carbon dioxide

Because you breathe in and out of the machine all the time, you can't get rid of your waste carbon dioxide in the normal way. There has to be a special filter to remove the carbon dioxide so it doesn't poison you!

- Carbon dioxide is a waste product of respiration, it is excreted through the lungs.
- Some of the amino acids we take in are not used. They are converted into urea by the liver and excreted by the kidneys in the urine. Urine can be stored in the bladder.
- The water and ion content of cells must be carefully controlled. If they are not, then too much or too little water may move in and out of cells by osmosis.

Key words: homeostasis, amino acids, urea, urine, liver, kidney, ions, osmosis

GET IT RIGHT!

Urea is produced by the liver and excreted by the kidneys. A lot of exam candidates get this the wrong way round (or forget it altogether!).

CHECK YOURSELF

1 Which process results in the production of carbon dioxide?

2 What are unwanted amino acids converted into?

3 Where is urine stored?

B2 5.2 Controlling body temperature

GET IT RIGHT!

Most students cannot explain how sweating cools you down. Remember that the evaporation of the sweat *takes energy from the skin* – that's why you cool down!

The thermoregulatory centre of the brain and receptors in the skin detect changes in temperature. The thermoregulatory centre controls the body's response to a change in internal temperature.

If the core temperature *rises*:

- Blood vessels near the surface of the skin dilate allowing more blood to flow through the skin capillaries. Heat is lost by radiation.
- Sweat glands produce more sweat. This evaporates from the skin's surface. The energy required for it to evaporate comes from the skin's surface. So we cool down.

If the core temperature *falls*:

- Blood vessels near the surface of the skin constrict and less blood flows through the skin capillaries. Less heat is radiated.
- We 'shiver'. Muscles contract quickly. This requires respiration and some of the energy produced is released as heat.

Key words: thermoregulatory, dilate, constrict, radiation, capillaries

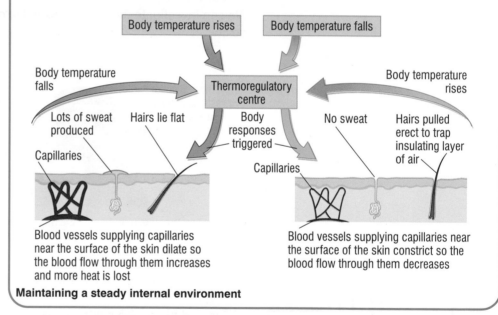

Maintaining a steady internal environment

CHECK YOURSELF

1 Which two parts of the body detect changes in temperature?

2 Where is sweat produced? [Higher Tier only]

3 Why does 'shivering' make you feel warmer? [Higher Tier only]

KEY POINTS

1 The level of sugar in the blood must be kept at the correct level.
2 Hormones help our bodies to do this.

The pancreas monitors and controls the level of sugar in our blood.

If there is too much sugar in our blood the pancreas produces the hormone insulin that results in the excess sugar being stored in the liver as glycogen. If insulin is not produced the blood sugar level may become fatally high.

If the pancreas is not producing enough insulin, this is known as diabetes. It can sometimes be controlled by diet or the person may need insulin injections.

Key words: pancreas, monitor, insulin, glycogen, diabetes

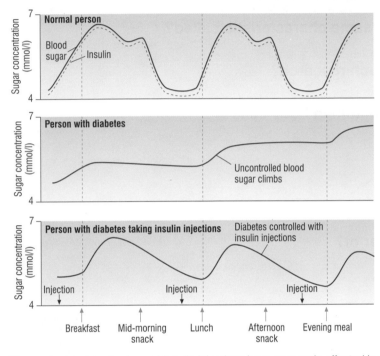

These graphs show the impact insulin injections have on people affected by diabetes. The injections keep the blood sugar level within safe limits.

AQA EXAMINER SAYS...

Remember it is when the blood sugar level is too high that insulin is produced.

CHECK YOURSELF

1 Which organ monitors blood sugar level?

2 Where is glycogen stored?

3 When the pancreas does not produce enough insulin, what is the condition known as?

B2 5   End of chapter questions

1 **Why is the ion content of the cells important?**

2 **What are the roles of the thermoregulatory centre?**

3 **Why does sweating cool you down? [Higher Tier only]**

4 **What happens if, after a large meal, there is too much sugar in the blood?**

5 **When your skin flushes, how is heat lost from the skin? [Higher Tier only]**

6 **How is urea eventually excreted?**

7 **What happens to the blood vessels near the surface of the skin if it is cold? [Higher Tier only]**

8 **If there is an increase in the concentration of substances in a living cell, which process will be affected?**

Pre Test: Inheritance

1. In which cells are chromosomes normally found in pairs?

2. When a cell divides by mitosis, how many new cells are formed?

3. Why are cells produced by mitosis?

4. How many cells are produced from a parent cell in meiosis? [Higher Tier only]

5. What are pairs of genes controlling the same characteristic called?

6. Why are stem cells the subject of so much research?

7. How many pairs of chromosomes are in human body cells?

8. What does DNA stand for?

9. What do we mean by a 'dominant' gene?

10. What is Huntingdon's disease a disorder of?

---

students' book page 202

B2 6.1 Cell division and growth

KEY POINTS

1 Body cells need to divide to produce new cells for growth or repair.
2 Mitosis is the type of cell division that produces identical new cells.

Cell division is necessary for the growth of an organism, or for repair if tissues are damaged.

Mitosis results in two identical cells being produced from the original cell.

A copy of each chromosome is made before the cell divides and one of each chromosome goes to each new cell.

Key words: mitosis, growth, repair

GET IT RIGHT!

Mitosis results in two new cells each identical to the parent cell. The new cells are either for growth or replacement (repair).

CHECK YOURSELF

1 Why do cells divide by mitosis?

2 Why is each new cell identical to the parent cell?

3 How many cells are produced in each division by mitosis?

B2 6.2 Stem cells

KEY POINTS

1 Stem cells are not specialised, but can differentiate into many different types of cell when required.
2 There are ethical issues surrounding the use of stem cells.

BUMP UP YOUR GRADE

There is a lot of argument about the use of embryonic stem cells for research. There are major ethical issues. You could be asked to offer your view and you may have a strong opinion. In the exam, give both sides of the argument if you want to gain full marks.

Stem cells are unspecialised. They can develop (differentiate) into many different types of specialised cell. Stem cells are found in the embryo and in adult bone marrow.

Many embryonic stem cells that we carry research out on are from aborted embryos, or are 'spare' embryos from fertility treatment. This results in ethical issues and much debate, as it can be argued that you are destroying life to obtain these stem cells for research.

The use of stem cells from adult bone marrow is still limited by the number of different types of specialised cell we can develop them into.

Key words: stem cell, specialised, embryonic, bone marrow

CHECK YOURSELF

1 Why are stem cells important?

2 Why are there ethical issues surrounding their use?

3 Why can we not just use stem cells from adult bone marrow?

B2 6.3 Cell division in sexual reproduction

HIGHER

KEY POINTS

1 Sex cells are produced by meiosis. [Higher Tier only]
2 Four cells are produced from each parent cell. They are all different. [Higher Tier only]

AQA EXAMINER SAYS...

Make sure that you can spell mitosis and meiosis. You may answer a question very well and lose nearly all of the marks, if the examiner cannot tell whether you are talking about mitosis or meiosis.

Cells in reproductive organs, e.g. testes and ovaries, divide to form sex cells (gametes).

Before division, a copy of each chromosome is made. The cell now divides twice to form four gametes (sex cells). This type of cell division is called meiosis.

Each gamete has only one chromosome from the original pair. All of the cells are different from each other and the parent cell.

Sexual reproduction results in variation as the sex cells (gametes) from each parent fuse. So half the genetic information comes from the father and half from the mother.

Key words: meiosis, gametes, variation

CHECK YOURSELF

1 How many sex cells are produced in one meiotic division? [Higher Tier only]

2 How many times does the parent cell divide to produce the gametes? [Higher Tier only]

3 Why does sexual reproduction result in variation?

B2 6.4 From Mendel to DNA

KEY POINTS

1 Gregor Mendel worked out how characteristics are inherited.
2 Genes make up the chromosomes, which control our characteristics.

Gregor Mendel was a monk who worked out how characteristics were inherited. His ideas were not accepted for many years.

Genes are short lengths of DNA (deoxyribonucleic acid), which make up chromosomes and control our characteristics.

Genes code for combinations of specific amino acids, which make up proteins.

HIGHER

Key words: genes, DNA, chromosomes

CHECK YOURSELF

1 What was the name of the monk who worked out the patterns of inheritance?

2 What does DNA stand for?

3 What are genes made up of?

B2 6.5 Inheritance in action

KEY POINTS

1 Alleles control the development of characteristics.
2 Some alleles are dominant and some are recessive.

- Human beings have 23 pairs of chromosomes, one pair are the sex chromosomes. Females are XX and males XY.
- Genes controlling the same characteristic are called alleles.
- If an allele 'masks' the effect of another it is said to be 'dominant'. The allele where the effect is 'masked' is said to be 'recessive'.

For example, the allele for brown eyes is dominant to the allele for blue eyes.

If two parents have brown eyes and have the genetic make up Bb, what would be the chance of them having a blue eyed child?

HIGHER

Parents	Bb	Bb
Sex cells	B or b	B or b

Fertilisation →

	B	b
B	BB	Bb
b	bB	**bb**

There is a 1 in 4 (25%) chance of having a blue eyed child (bb)

Key words: allele, dominant, recessive

Sex chromosomes

The chromosomes of the human male

AQA EXAMINER SAYS...

You can draw Punnett Squares to show genetic crosses or 'line' diagrams. It is easier to make mistakes with 'line' diagrams and also harder for the examiner to see what you are doing.

CHECK YOURSELF

1 How many chromosomes does a human being have?

2 What are the sex chromosomes of a male?

3 What is meant by a 'recessive allele'?

HIGHER

Huntington's disease is a disorder of the nervous system. It is caused by a dominant allele, so even if only one parent has the disease it can be inherited by a child.

Cystic fibrosis is a disorder of cell membranes. It is caused by a recessive allele so parents may be carriers (Cc). Only if both parents are either carriers or have the disorder does a child inherit it.

Embryos can be screened to see if they carry alleles for one of these or other genetic disorders.

Key words: disorder, Huntington's, cystic fibrosis, carrier

Both parents are carriers, so Cc

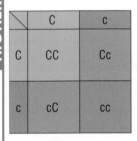

	C	c
C	CC	Cc
c	cC	cc

25% normal (CC)
50% carriers (Cc)
25% affected by cystic fibrosis (cc)

3/4, or 75% chance normal
1/4, or 25% chance cystic fibrosis

The faulty alleles can be covered up by normal alleles for generations until two carriers have a child and, by chance, both of the cystic fibrosis alleles are passed on

Parent with Huntington's disease   Hh
Normal parent   hh

	H	h
h	Hh	hh
h	hH	hh

50% chance Huntington's disease, Hh or hH
50% chance normal, hh

A genetic diagram for Huntington's disease shows how a dominant allele can affect offspring

B2 6   End of chapter questions

1 When we say that stem cells 'differentiate', what do we mean?

2 What is the result of one meiotic cell division? [Higher Tier only]

3 What is meant by 'dominant' when we refer to alleles?

4 What is a 'gene'?

5 What do the chromosomes do before cell division takes place?

6 How many pairs of chromosomes are present in human cells (other than sex cells)?

7 What are 'alleles'?

8 What is cystic fibrosis a disorder of?

1 We need to keep the internal conditions in our bodies constant. The diagram shows some of the organs that do this.

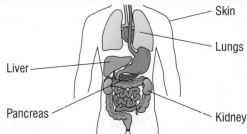

(a) Which of the organs in the diagram:
  (i) excretes carbon dioxide
  (ii) produces urea
  (iii) excretes urea in the urine
  (iv) produces sweat? (4 marks)

(b) How does sweat help to cool the body down on a hot day? (3 marks)

(c) On a cool day we might 'shiver'. How does 'shivering' help us to become warmer? (2 marks) [Higher]

2 (a) Where are stem cells found? (2 marks)

(b) Why is research into stem cells so important? (2 marks)

(c) Why is there so much argument about whether research into stem cell research should be taking place at all? (3 marks)

3 Meat is protein. Proteins are digested by protease enzymes.

A student investigated the effect of pH on the digestion of cubes of meat by protease enzymes. The student kept all of the other variables the same. The table shows the student's results.

pH	Time taken to digest the meat (minutes)
1	14
2	12
3	18
4	32
5	52
6	lesson had ended

(a) (i) What pH provided the best conditions for the enzyme to work in? (1 mark)
  (ii) From the information given above, how could the student have improved the investigation? (1 mark)

(iii) What further work could the student carry out to find out more accurately the best pH for this enzyme? (2 marks)

(b) In which part of the gut do you think this protease enzyme is found? Explain your answer. (2 marks)

(c) (i) What do the following enzymes digest and what are the products of this digestion?
    • lipase   • amylase   (5 marks)
  (ii) Where is the enzyme amylase produced? (2 marks)

(d) What is the function (job) of bile? (2 marks)

4 It is very important that we keep our blood sugar at a fairly constant level in the body.

(a) Why do we need sugar in the blood? (1 mark)

(b) Which organ monitors the blood sugar level? (1 mark)

(c) How is blood sugar brought back down to the 'normal' level if it becomes too high after a meal? (3 marks)

(d) If you cannot control the blood sugar level in the body you may suffer from diabetes. How is diabetes controlled? (2 marks)

5 Huntington's disease is a rare disorder of the nervous system. It is inherited through a dominant allele represented by H. The recessive allele of this pair of genes is represented by h.

Study the diagram below, which shows the inheritance of Huntington's disease in a family.

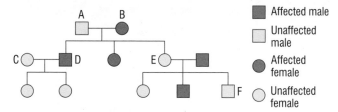

(a) Use a genetic diagram to show why D has Huntington's disease. (3 marks)

(b) Why does E not have the disease? (2 marks)

(c) If F were to have children with a person carrying one allele for Huntington's disease, what would be the chance of a child having the disease? Explain your answer. (2 marks)

(d) (i) How does DNA control the inheritance of characteristics of organisms? (2 marks)
  (ii) What is a short length of DNA coding for one characteristic called? (1 mark) [Higher]

 Test & Assessment Interactive quizzes, answers and hints online!

This is a very difficult question (Higher Tier only) and the student has gained 3 of the 4 marks available.

The responses worth a mark are underlined in red.

We can improve the answer in several ways:

In meiosis the chromosomes do copy, but don't 'split in half'.

Many sex cells are produced and which sperm fertilises which egg is a random process. The sex cells will carry the genes from each parent and the candidate is correct in that dominant or recessive alleles may be carried.

The sex cells are produced by meiosis.
Describe what happens to the chromosomes during meiosis and how meiosis and sexual reproduction lead to variation in the offspring. *(4 marks)*

During meiosis each chromosome splits itself in half and copies each half. They then all line up in the middle of the cell and are separated by strands from the cell. In the new cells they double again and repeat the process.

During fertilisation we do not know which of the sex cells (copies) will fuse together. Each instruction is taken from the male and female chromosome, even then the alleles could be dominant or recessive.

Then one from each pair goes to each of the two new cells.

The answer is worth 4 marks out of the 6 marks available.

The responses worth a mark are underlined in red.

The student has used the information well and gains all 4 of the marks for the first section. Most students should be able to use information presented in this way.

In the second part the student scores no marks.

We can improve the answer in several ways:

We use enzymes in industry. Here are some properties of enzymes.
- **They work at low temperatures so can save energy and, therefore, money.**
- **Some reactions require a high pressure, enzymes lower the pressure necessary.**
- **They are easily broken down at too high a temperature or the wrong pH.**
- **They are soluble in water, so are difficult to separate when a reaction has finished.**
- **They are produced from microorganisms, so are expensive to buy.**
(a) Give two advantages and two disadvantages of using enzymes in industry? *(4 marks)*
(b) Why don't enzymes work at high temperatures? *(2 marks)*

(a) Advantages are they work at low temperatures so save money and low pressure so you don't need such expensive equipment. Disadvantages are that they cost a lot and it is difficult to get them back after the reaction.
(b) They don't work at high temperatures as they are killed.

The enzymes don't work as their shape is altered. They are therefore denatured (or destroyed). Never say that enzymes are killed – they are chemicals!

B3 | Further biology (Chapters 1–2)

Checklist

This spider diagram shows the topics in the unit. You can copy it out and add your notes and questions around it, or cross off each section when you feel confident you know it for your exams.

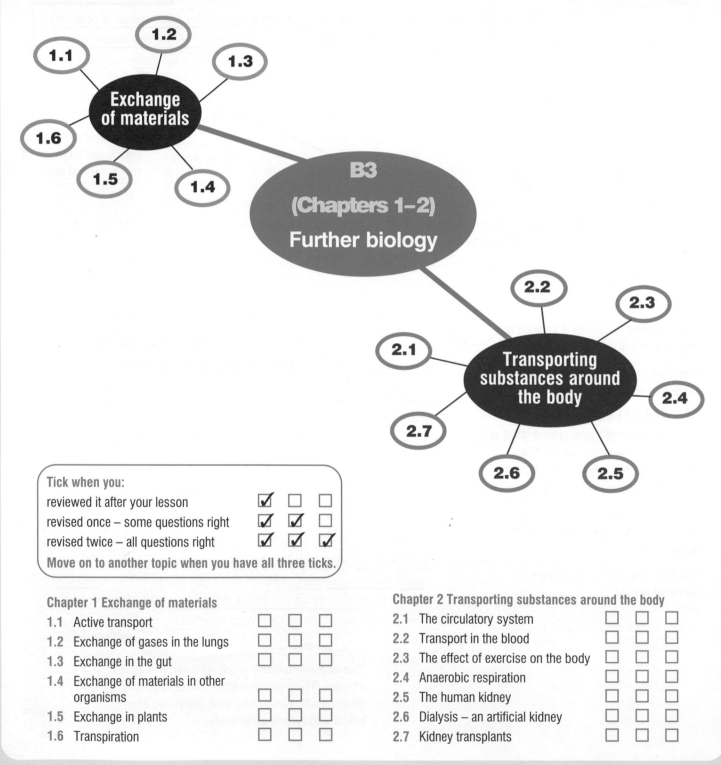

Tick when you:

reviewed it after your lesson	✓	☐	☐
revised once – some questions right	✓	✓	☐
revised twice – all questions right	✓	✓	✓

Move on to another topic when you have all three ticks.

Chapter 1 Exchange of materials

1.1	Active transport	☐	☐	☐
1.2	Exchange of gases in the lungs	☐	☐	☐
1.3	Exchange in the gut	☐	☐	☐
1.4	Exchange of materials in other organisms	☐	☐	☐
1.5	Exchange in plants	☐	☐	☐
1.6	Transpiration	☐	☐	☐

Chapter 2 Transporting substances around the body

2.1	The circulatory system	☐	☐	☐
2.2	Transport in the blood	☐	☐	☐
2.3	The effect of exercise on the body	☐	☐	☐
2.4	Anaerobic respiration	☐	☐	☐
2.5	The human kidney	☐	☐	☐
2.6	Dialysis – an artificial kidney	☐	☐	☐
2.7	Kidney transplants	☐	☐	☐

What are you expected to know?

Chapter 1 Exchange of materials (See students' book pages 222–235)

- How active transport in cells differs from diffusion and osmosis. [Higher Tier only]

- The lungs and the small intestine have very large surface areas for the exchange of substances.

- The lungs have millions of air sacs called alveoli.

- Carbon dioxide and oxygen are exchanged by the lungs.

- The blood absorbs the products of digestion from the small intestine.

- Thousands of villi increase the surface area of the small intestine so that absorption takes place.

- Plants exchange carbon dioxide and oxygen through the surface of the leaves.

- Plants take water and minerals up through the roots.

- Root hairs greatly increase the surface area of roots.

- Water is lost by plants through transpiration.

- Temperature, humidity and wind speed all affect the rate of transpiration.

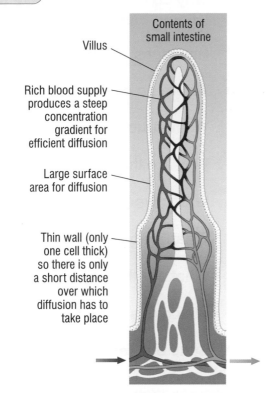

Contents of small intestine

Villus

Rich blood supply produces a steep concentration gradient for efficient diffusion

Large surface area for diffusion

Thin wall (only one cell thick) so there is only a short distance over which diffusion has to take place

Chapter 2 Transporting substances around the body (See students' book pages 238–253)

- In humans there is a double circulatory system – one is to the lungs, the other is to the body.

- Red blood cells transport oxygen in the blood – oxygen attaches to haemoglobin to form oxy-haemoglobin.

- Blood plasma transports a number of other substances.

- During exercise the heart and breathing rate are both increased and some blood vessels dilate.

- The energy needed by muscles to contract is usually provided through aerobic respiration.

- If insufficient oxygen is transported to the muscles then they may respire anaerobically.

- Anaerobic respiration is inefficient and the muscles will become fatigued.

- The kidneys remove certain substances from the body including urea and excess water.

- If the kidneys fail, then dialysis or kidney transplants can keep the person alive.

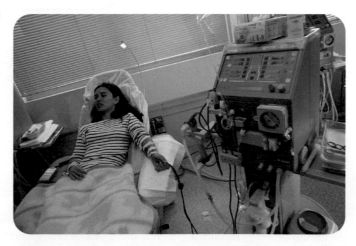

Pre Test: Exchange of materials

1. Why do cells need to take up some substances by 'active transport'? [Higher Tier only]

2. How does active transport differ from diffusion and osmosis? [Higher Tier only]

3. Where do cells produce most of their energy?

4. What structures increase the surface area of the lungs?

5. Why are the surfaces of the lungs moist?

6. Which type of blood vessel picks up oxygen at the lungs?

7. What is absorbed by the blood from the small intestine?

8. What are the tiny holes in the surface of a leaf called?

9. Which gas diffuses out of leaves during darkness?

10. State one factor that will increase the rate of transpiration.

---

students' book page 222

B3 1.1 Active transport

KEY POINTS

Substances get in and out of cells through:
1 diffusion
2 osmosis
3 active transport.

Osmosis and diffusion involve movement of substances from a higher concentration to a lower concentration.

HIGHER

A cell may need to take up a substance against the concentration gradient. This involves energy and is called 'active transport'.

Key words: concentration gradient, energy, active transport

EXAMINER SAYS...

Active transport is the only one of the three methods that requires energy. It is only worth the cell using this energy if the substance involved is really needed.

Mineral ions moved into plant **against** a concentration gradient

Mineral ions in soil – low concentration

Mineral ions in plant – higher concentration

It takes the use of energy in active transport to move mineral ions against a concentration gradient like this

CHECK YOURSELF

1 Why do diffusion and osmosis not need energy?

2 What do we mean by the term 'concentration gradient?' [Higher Tier only]

3 Why does active transport require energy? [Higher Tier only]

B3 1.2 Exchange of gases in the lungs

KEY POINTS

1 The lungs are in the thorax and are protected by the rib cage.
2 The lungs exchange carbon dioxide and oxygen with the atmosphere.
3 They have a large surface area and are moist.

GET IT RIGHT!

The surfaces of the lungs are moist as diffusion takes place much more quickly through a wet surface (in water).

The lungs exchange oxygen (needed for aerobic respiration) and carbon dioxide (a waste product of respiration) with the atmosphere. They have a very large surface area provided by millions of air sacs (alveoli). The surfaces of the lungs are moist and thin so that diffusion takes place quickly.

Oxygen diffuses into the many capillaries surrounding the alveoli and carbon dioxide diffuses back out into the lungs.

Key words: diffuses, alveoli, surface area, moist

CHECK YOURSELF

1 Which gas is a waste product of respiration?
2 Why do the lungs have a very large surface area?
3 Which type of blood vessel picks up oxygen at air sacs?

B3 1.3 Exchange in the gut

KEY POINTS

1 Digested food is absorbed by the capillaries alongside the small intestine (gut).
2 The small intestine has a 'rich' blood supply.
3 The surface area of the small intestine is greatly increased by villi.
4 Absorption by the blood is by diffusion and active transport.

Food we eat is digested in the gut into small, soluble molecules. In the small intestine these are absorbed by the blood. Finger-like projections into the small intestine, called villi, greatly increase the surface area for absorption to take place.

Food is absorbed by diffusion, where there is a lower concentration of that molecule in the blood. Food is absorbed by active transport where movement is against the concentration gradient.

Key word: villi (villus: singular)

Structure of the small intestine
— Villus
— Rich blood supply produces a steep concentration gradient for efficient diffusion
— Large surface area for diffusion
— Thin wall (only one cell thick) so there is only a short distance over which diffusion has to take place

Villi make it possible for all the digested food molecules to be transferred from your small intestine into your blood by diffusion

CHECK YOURSELF

1 Where does the absorption of digested food take place?
2 Which structures increase the surface area of the small intestine?
3 Which two processes result in digested food entering the bloodstream from the small intestine?

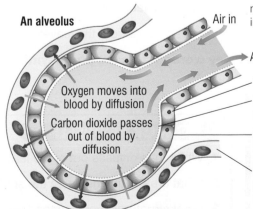

An alveolus

Air in

Oxygen moves into blood by diffusion

Carbon dioxide passes out of blood by diffusion

Air out
— Very thin walls make diffusion easy
— Moist surface makes diffusion easy as gases can dissolve
— Spherical shape gives relatively large surface area for diffusion
— Good blood supply maintains concentration gradient for diffusion by removing oxygen and bringing lots of carbon dioxide

The alveoli are adapted so that gas exchange can take place as efficiently as possible in the lungs

B3 1.4 Exchange of materials in other organisms

KEY POINTS

1 Fish exchange oxygen through their gills.
2 Frogs exchange oxygen through their skin.
3 Insects exchange oxygen through holes in their sides leading to a series of tubes.

EXAMINER SAYS...

These ideas about exchange surfaces are very important. You need to know them for the examination. However, you do not need to know any detail about the organisms mentioned. If there are any questions relating to fish, frogs or insects you will be given information and asked to apply your knowledge.

All living organisms need to exchange gases. They need oxygen for respiration and to remove carbon dioxide (the waste product of respiration). these organisms have a number of features in common:

● They have a large surface area.
● They are moist.
● The gases are transported away quickly to maintain a high concentration gradient.
● The membranes which the gases diffuse across are thin.

Key words: surface area, concentration gradient, membrane

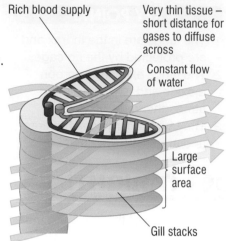

Rich blood supply

Very thin tissue – short distance for gases to diffuse across

Constant flow of water

Large surface area

Gill stacks

The gills of a fish

CHECK YOURSELF

1 Why do exchange surfaces have large surface areas?
2 Why are surface areas nearly always moist?
3 Why does a high concentration gradient need to be maintained?

B3 1.5 Exchange in plants

KEY POINTS

1 Plants exchange gases through their leaves.
2 Plants take up water and minerals through their roots.
3 Leaves and roots have adaptations for efficient exchange of materials.

GET IT RIGHT!

Plants respire all of the time. This requires oxygen.
● During daylight hours enough oxygen is produced by photosynthesis, so diffusion into the leaves is mainly of carbon dioxide.
● At night oxygen diffuses into the leaves, as respiration is the only process taking place.

Gases diffuse in and out of leaves through tiny holes called 'stomata'.

● Oxygen is needed for respiration and is a waste product of photosynthesis.
● Carbon dioxide is needed for photosynthesis and is a waste product of respiration. The movement of these gases depends upon which process is taking place the most quickly.

Leaves are flat and very thin so the gases do not need to diffuse very far. There are also internal air spaces.

Water and mineral ions are taken up by the roots. Roots have thousands of tiny projections called root hairs to increase their surface area. (See next page.)

Water evaporates from the leaves. This can be a problem.

Key words: stomata, photosynthesis, respiration, diffusion

CHECK YOURSELF

1 Which gas diffuses out of the leaves during photosynthesis?
2 What are the tiny holes on the surface of leaves called?
3 What structures increase the surface area of the roots?

B3 1.6 Transplantation

Plants take up water through the roots. The water passes through the plant to the leaves. It evaporates from the leaves in a process known as transpiration.

1 Plants take up gases through stomata.
2 They take up water and can lose water through the stomata.
3 If plants lose too much water they wilt.

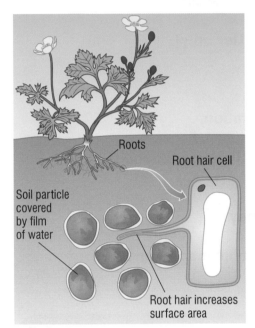

Many small roots, and the presence of microscopic root hairs on the individual root cells all increase diffusion of substances from the soil into the plant

Tiny holes called 'stomata', usually on the underside of the leaf, allow gases to be exchanged but also water to be lost.

If too much water is lost then the plant will wilt. Guard cells control the size of the stomata. The stomata can close to prevent water loss. More water is lost on hot, windy and dry days.

Key words: transpiration, evaporate, stomata, guard cells, wilt

 **EXAMINER SAYS...**

These ideas about exchange surfaces are very important, you need to know them for the examination. However, you do not need to know any detail about the organisms mentioned. If there are any questions relating to fish, frogs or insects you will be given information and asked to apply your knowledge.

GET IT RIGHT!

Most candidates can remember that transpiration takes place more quickly on hot, dry and windy days. Only candidates achieving the highest grades understand why:

● On hot days ... there is more energy causing the water to evaporate.
● On dry days ... the air can hold more water.
● On windy days ... any build up of humidity (water vapour) around the plant is blown away.

CHECK YOURSELF

1 Why do stomata need to be open at least some of the time?
2 Which cells open and close the stomata?
3 Why do you think that stomata are usually found on the lower surface of leaves?

B3 1 End of chapter questions

1 What do we mean when we say particles 'move against the concentration gradient'? [Higher Tier only]

2 Which process releases the energy for 'active transport'? [Higher Tier only]

3 Why are the surfaces of the lungs thin?

4 Why is it necessary for the small intestine to have a 'rich' blood supply?

5 What feature do all exchange surfaces in humans have in common?

6 Why are leaves very thin?

7 Why do humid days result in less transpiration from leaves, even if it is hot?

8 What substances do roots take up?

Pre Test: Transporting substances around the body

1. Which type of blood vessel exchanges materials with the cells?

2. Which organ does the blood transport urea to?

3. What is the name of the red pigment in red blood cells?

4. Why do muscles need more blood flowing to them during exercise?

5. What are the products of anaerobic respiration? [Higher Tier only]

6. What do we mean by 'oxygen debt'? [Higher Tier only]

7. Name two substances that the kidney reabsorbs after filtration.

8. Why might energy be required for the kidney to reabsorb these substances? [Higher Tier only]

9. Which system in the body may reject a transplanted kidney?

10. Other than a kidney transplant, what other option is available if someone's kidneys fail?

---

students' book page 238

B3 2.1 The circulatory system

KEY POINTS

1. The heart pumps blood around the body.
2. There are two separate circulation systems.
3. Arteries, veins and capillaries are the vessels involved in transporting the blood.

AQA EXAMINER SAYS...

It is only the capillaries that allow the exchange of substances with the tissues and organs. The walls of arteries and veins are much too thick for substances to pass through.

The heart is a muscular organ that pumps blood around the body. The heart pumps blood to the lungs where it picks up oxygen and loses carbon dioxide. After returning to the heart it is then pumped to the rest of the body. These are the two circulations.

Arteries take blood away from the heart and the blood returns in veins. Capillaries are very small vessels that are between the arteries and veins. They carry the blood through the organs and allow the exchange of substances with all the living cells in the body.

Key words: heart, artery, vein, capillary, circulation

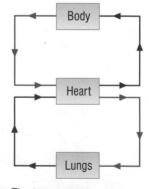

The two separate circulation systems supply the lungs and the rest of the body

CHECK YOURSELF

1. How is the heart able to pump blood around the body?

2. Which blood vessels carry blood away from the heart?

3. Which gas is lost from the blood at the lungs?

B3 2.2 Transport in the blood

GET IT RIGHT!

The red blood cells only transport oxygen – try to remember how to spell haemoglobin! All of the other substances are transported in solution in the plasma.

CHECK YOURSELF

1 Which organs is urea transported to?

2 Where does the oxyhaemoglobin release its oxygen?

3 Why do red blood cells have no nucleus?

Red blood cells are packed with a red pigment called 'haemoglobin'. In the lungs, haemoglobin combines with oxygen to form oxyhaemoglobin.

Where the cells are respiring, the oxyhaemoglobin breaks back down releasing the oxygen for respiration. Red blood cells have no nucleus so that more haemoglobin can be packed into them.

The blood plasma transports:

● carbon dioxide to the lungs,
● the (soluble) products of digestion to all living cells in the body, and
● urea, made by the liver, to the kidneys where it is excreted.

Key words: haemoglobin, oxyhaemoglobin, plasma

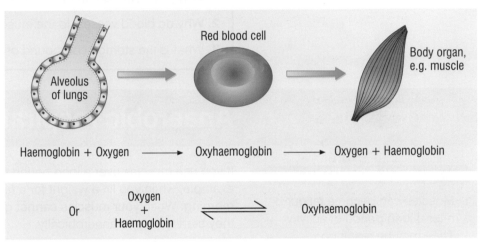

This reversible reaction makes active life as we know it possible by carrying oxygen to all the places where it is really needed

B3 2.3 The effect of exercise on the body

When you exercise your muscles need more energy so that they can contract. You need to increase the rate at which oxygen and glucose reach the muscle cells. You also need to remove carbon dioxide more quickly.

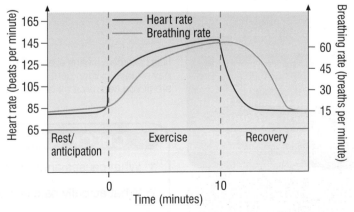

During exercise the heart rate and breathing rate increase to supply the muscles with what they need and remove the extra waste produced

Many candidates will remember that if you exercise you will need more energy. Fewer students will remember that you will, therefore, need more oxygen and glucose. Very few indeed will write about the need to excrete more carbon dioxide.

The heart rate increases and the blood vessels supplying the muscles dilate. This allows more blood containing oxygen and glucose to reach the muscles.

Your breathing rate and the depth of each breath also increase. This allows a greater take-up of oxygen and release of carbon dioxide at the lungs.

You may also use up glycogen, which is a storage compound of glucose, present in the muscles.

Key words: rate, glycogen

CHECK YOURSELF

1 Why does your depth of breathing increase during exercise?

2 Why do blood vessels to the muscles dilate while exercising?

3 What is the storage compound of glucose?

students' book
page 244

B3 2.4   Anaerobic respiration

KEY POINTS

1 Muscles can become fatigued after a long period of activity. They may be respiring anaerobically as they cannot get enough oxygen.

2 Lactic acid is produced in anaerobic respiration. When the activity has stopped lactic acid must be broken down. This is the cause of 'oxygen debt'. [Higher Tier only]

If you use muscles over a long period then they will get tired or fatigued. For example, when you lift a weight for a few minutes or run for the school bus in the morning. When your muscles cannot get enough oxygen for aerobic respiration, they start to respire anaerobically.

This is inefficient and produces lactic acid as a waste product.

Lactic acid causes fatigue. When the exercise has finished this lactic acid must be completely broken down. You therefore still need to take in a lot of oxygen to do this. This is known as 'oxygen debt'. The oxygen oxidises lactic acid into carbon dioxide and water.

Key words: fatigue, anaerobic, lactic acid, oxygen debt

Remember that it is the lactic acid that causes fatigue (tiredness). When you stop your exercise you must get rid of this lactic acid build-up. You keep breathing in a lot of oxygen to do this. Warming down after exercise helps to get rid of the lactic acid.

Hard exercise means everyone has to pay off their oxygen debt – but if you are fit you can pay it off faster!

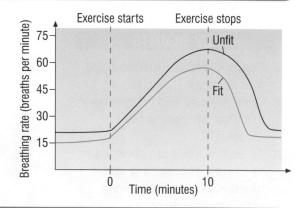

CHECK YOURSELF

1 What causes anaerobic respiration to take place?

2 What actually causes your muscles to feel fatigued? [Higher Tier only]

3 What is lactic acid finally broken down into? [Higher Tier only]

B3 2.5 The human kidney

KEY POINTS

1 Chemical reactions in the body produce substances that are toxic (poisonous) e.g. urea.
2 The kidneys excrete substances that the body does not want.
3 The kidneys first filter substances out of the blood. They then reabsorb the substances that the body needs.

GET IT RIGHT!

Urea is made by the liver from excess amino acids. Urine is the liquid excreted by the kidneys which has urea dissolved in it. It is also important to know the difference between filtration and reabsorption.

The body has two kidneys. They filter the blood, excreting substances you do not want and keeping those substances that the body needs.

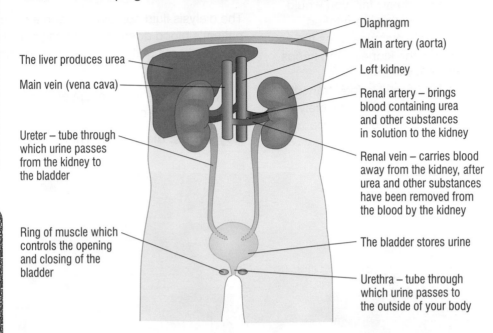

- Diaphragm
- Main artery (aorta)
- Left kidney
- Renal artery – brings blood containing urea and other substances in solution to the kidney
- Renal vein – carries blood away from the kidney, after urea and other substances have been removed from the blood by the kidney
- The bladder stores urine
- Urethra – tube through which urine passes to the outside of your body

- The liver produces urea
- Main vein (vena cava)
- Ureter – tube through which urine passes from the kidney to the bladder
- Ring of muscle which controls the opening and closing of the bladder

The kidney, a very important organ of homeostasis, is involved in controlling the loss of water and mineral ions from the body as well as getting rid of urea. The kidneys have a very rich blood supply.

After filtering the blood, all of the sugar and many of the dissolved ions and water needed by the body are reabsorbed. All of the urea and some ions dissolve in the remaining water and are excreted in the urine.

Sugar and dissolved ions may be reabsorbed against the concentration gradient. If they are, then the process is active transport and requires energy.

Key words: filter, reabsorb, excrete, active transport

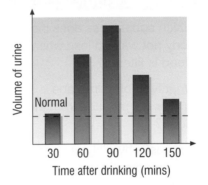

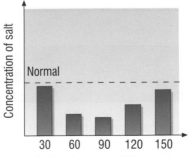

These data show how your kidneys respond when you drink a lot. They show volume of urine produced and the concentration of salt in the urine after a student drank a large volume of water.

CHECK YOURSELF

1 Which substance, after filtration, is the only one not reabsorbed at all by the kidneys?

2 Why might the process of reabsorbing glucose require energy? [Higher Tier only]

3 Are you more likely to produce more urine on a warm day or on a cold day? Assume that you drink the same amount.

HIGHER

B3 2.6 Dialysis – an artificial kidney

KEY POINTS

1 If your kidneys fail, you would die if no action were taken.
2 A dialysis machine does the work of the kidneys and keeps patients alive.
3 If a successful kidney transplant is carried out, then the dialysis machine will no longer be necessary.

A dialysis machine carries out the same job as the kidneys. The blood flows between partially permeable membranes.

The dialysis fluid contains the same concentration of useful substances that the patient's blood does, e.g. glucose and mineral ions. This means that these are not filtered out of the blood so they do not need to be reabsorbed. Urea does get filtered out.

Dialysis restores the concentration of substances in the blood back to normal but needs to be carried out at regular intervals.

Key words: dialysis, partially permeable membrane, restores, urea

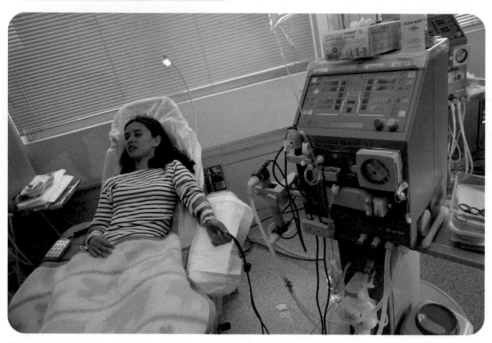

These 'artificial kidney machines' have not only saved countless lives, but allowed sufferers from kidney failure to lead full, active lives

GET IT RIGHT!

In dialysis the useful substances in the blood are not filtered out so there is no need for a system to reabsorb them. The partially permeable membranes in the machine do the same job as the cell membranes in the kidney.

CHECK YOURSELF

1 Which substances does the body not want to lose when the blood is filtered?

2 What type of membrane is used to filter out the urea from the blood?

3 Suggest why dialysis needs to be carried out at regular intervals.

B3 2.7 Kidney transplants

KEY POINTS

1 A diseased kidney can be replaced by a healthy kidney.
2 The healthy kidney must be a very good 'match'.
3 The immune system has to be 'suppressed' or it is likely to reject the new kidney.

For most patients a kidney transplant is a better option than dialysis.

A donor kidney must be found, often from a person that has just died. The new kidney must be a very good 'tissue match'.

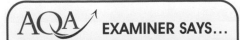
The immune system is likely to try to reject the new kidney so:

- The patient's (recipient's) bone marrow is treated with radiation to stop white blood cell production.
- The recipient will have to continue to take drugs to suppress the immune system (immunosuppressant drugs).
- After the operation, the recipient must be kept in sterile conditions to prevent infection.

Key words: donor, recipient, tissue match, immune, immunosuppressant

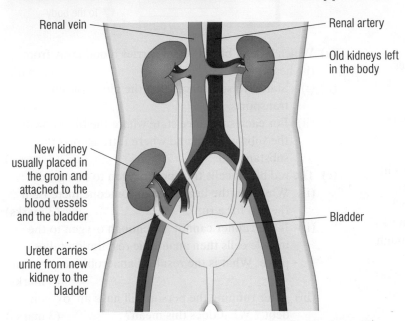

Renal vein

Renal artery

Old kidneys left in the body

New kidney usually placed in the groin and attached to the blood vessels and the bladder

Ureter carries urine from new kidney to the bladder

Bladder

A donor kidney is placed in the body where it takes over the functions of the organs which have failed

CHECK YOURSELF

1 Why is the production of white blood cells stopped when a patient is about to receive a new kidney?

2 Suggest why the new kidney must be as close a match as possible to the patient's tissue type.

3 Where do most donor kidneys come from?

B3 2   End of chapter questions

1 **Which type of blood vessel takes blood away from the heart?**

2 **How do red blood cells transport oxygen?**

3 **Which process carried out by living cells requires oxygen?**

4 **Why does the heart rate increase during exercise?**

5 **What is the waste product of anaerobic respiration? [Higher Tier only]**

6 **What causes oxygen debt? [Higher Tier only]**

7 **Which substances are found in urine?**

8 **Why does the glucose not filter out of the blood during dialysis?**

1 The diagram shows an alveolus in the lung.

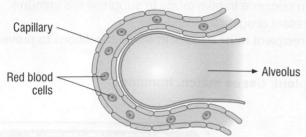

Capillary

Alveolus

Red blood cells

(a) State two features of the lung that help gases to exchange more efficiently with the blood capillaries. (2 marks)

(b) How do the red blood cells transport oxygen to the living cells in the body? (2 marks)

(c) The soluble products of digestion are absorbed by capillaries surrounding the small intestine.

(i) What is the name of the 'finger-like' structures in the small intestine that help the capillaries to absorb the products of digestion more quickly? (1 mark)

(ii) Some of the products of digestion diffuse through the small intestine wall. What is meant by the word 'diffusion'? (3 marks)

(iii) Some of the other products of digestion have to be taken up through the wall of the small intestine by 'active uptake'. What is meant by 'active uptake'? (2 marks)

[Higher]

2 The diagram shows a stoma found on the lower surface of a leaf.

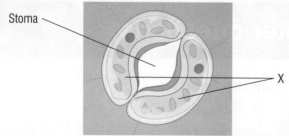

Stoma

X

(a) (i) What is the name of the cells labelled X? (1 mark)

(ii) What is the function (job) of these cells? (1 mark)

(b) Describe one adaptation of a leaf that helps gases to diffuse more quickly to the cells. (1 mark)

(c) Which gas would you expect to be diffusing into a leaf on a sunny day? Explain your answer. (2 marks)

(d) (i) Water is lost from leaves during the day by evaporation. What name do we give to this process in plants? (1 mark)

(ii) State two conditions that will increase the rate of evaporation. For each one explain why the plant would lose more water. (4 marks)

3 The diagram shows the two circulation systems in the human body.

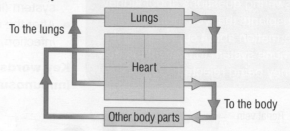

To the lungs

Lungs

Heart

To the body

Other body parts

(a) Which type of blood vessel carries blood away from the heart? (1 mark)

(b) (i) State two substances that the blood plasma transports. (2 marks)

(ii) For each substance state where the blood picks the substance up and where it transports the substance to. (4 marks)

(c) The red blood cells transport oxygen to living cells.

(i) Why does the heart rate of someone increase if they go for a run? (3 marks)

(ii) If the runner cannot get enough oxygen to the muscle cells then anaerobic respiration takes place. What is the result of anaerobic respiration? (3 marks)

(iii) After running the person will have an 'oxygen debt'? What does this mean? (3 marks)

[Higher]

4 The diagram shows one kidney tubule.

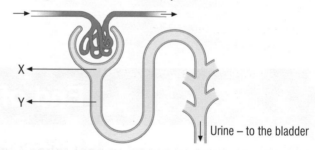

X

Y

Urine – to the bladder

(a) What process is happening at X? (1 mark)

(b) (i) Which substance is totally reabsorbed at Y? (1 mark)

(ii) Name one substance that is only partially reabsorbed at Y. (1 mark)

(c) (i) What type of membrane is used in a dialysis machine? (1 mark)

(ii) The dialysis fluid contains the same concentration of useful substances as the blood. Why is this necessary? (2 marks)

(d) It is cheaper for the health service for people with kidney failure to have a kidney transplant rather than to have dialysis. However some kidneys are rejected after being transplanted. Why are more people not offered kidney transplants? (2 marks)

 Test & Assessment Interactive quizzes, answers and hints online!

EXAMPLES OF EXAM ANSWERS

The answer is worth 2 marks out of the 4 marks available.

The responses worth a mark are underlined in red.

We can improve the answer in several ways:

The donor kidney should also be a close **tissue type** to the diseased kidney to further reduce the chance of rejection (it is not enough to state that it should be 'as much the same').

A transplant operation can be performed to replace a diseased kidney with a healthy kidney. It is very likely that the new kidney will be rejected by the body unless certain precautions are taken before and after the operation. Describe as fully as you can what these precautions are. *(4 marks)*

The donor kidney used has to be as much the same as the kidney that has become diseased to stop the <u>white blood cells</u> from attacking it. It is best if the white cells are stopped from attacking it. After the operation has finished the patient is kept in intensive care in a <u>sterile area</u>.

The white cells are part of the **immune system**. If the **bone marrow** is treated with **radiation** before the operation then the **white blood cell production will be stopped** (it is not good enough just to state that white blood cell production must be stopped, you need to say how).

The answer is worth 2 marks out of the 4 marks available.

The responses worth a mark are underlined.

We can improve the answer in several ways:

The diagram shows the lungs and some other structures in the thorax.

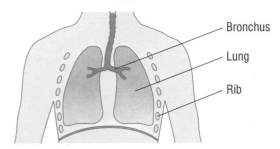

Bronchus

Lung

Rib

Which gases are exchanged in the lungs and why do they need to be exchanged? *(4 marks)*

<u>Carbon dioxide</u> and <u>oxygen</u> are changed in the lungs as you need to breathe them in and out. The body needs oxygen to keep the cells alive. It is important you keep breathing to keep the gases moving in and out.

Carbon dioxide is the **waste product** of respiration and needs to be breathed out or it would rise to **harmful levels**.

Oxygen is needed for **respiration / to release energy**. Oxygen needs to be **taken into** the blood.

We need to state why the gases need to be exchanged. This answer is too vague simply stating 'to keep the cells alive'.

Further biology (Chapter 3)

Checklist

This spider diagram shows the topics in the unit. You can copy it out and add your notes and questions around it, or cross off each section when you feel confident you know it for your exams.

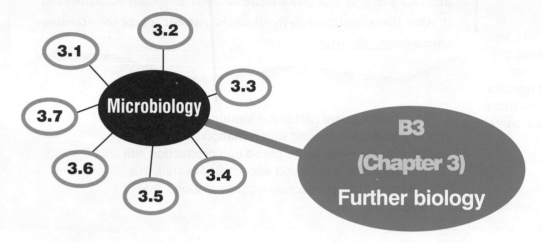

Tick when you:

reviewed it after your lesson	☑	☐	☐
revised once – some questions right	☑	☑	☐
revised twice – all questions right	☑	☑	☑

Move on to another topic when you have all three ticks.

Chapter 3 Microbiology

3.1	Growing microbes	☐	☐	☐
3.2	Food production using yeast	☐	☐	☐
3.3	Food production using bacteria	☐	☐	☐
3.4	Large-scale microbe production	☐	☐	☐
3.5	Antibiotic production	☐	☐	☐
3.6	Biogas	☐	☐	☐
3.7	More biofuels	☐	☐	☐

What are you expected to know?

Chapter 3 Microbiology (See students' book pages 256–271)

- Bacteria are used to make yoghurt and cheese.

- Yeast is used to make bread and alcohol (including beer and wine).

- Fermenters are large vessels used to grow microorganisms on a large scale.

- The conditions inside a fermenter are carefully controlled.

- Antibiotics are made from the mould *Penicillium* in fermenters.

- The fungus *Fusarium* is used to make mycoprotein.

- Biogas, which is mainly methane, can be made from the fermentation of 'natural' materials.

- Biogas generators can be used to generate gas on a large scale or a small scale.

- Ethanol-based fuels can be made from the fermentation of glucose from a variety of plant material.

- On a small scale, microorganisms can be grown on agar in Petri dishes.

- Agar is a 'culture medium' and contains carbohydrate and a variety of other growth substances.

- To maintain sterile conditions, care must be taken when growing microorganisms.

- In schools and colleges, and elsewhere, special care has to be taken to prevent the growth of harmful organisms (pathogens).

1. How are bacteria used in food manufacture?

2. How is yeast used in food manufacture?

3. What is the product of the anaerobic respiration of yeast?

4. In beer making what is the energy source for the yeast?

5. What is the name of the type of sugar found in milk?

6. Which fungus produces penicillin?

7. What is a 'biogas' generator?

8. Which new fuel, produced from sugars, is being used to replace petrol?

9. What is 'agar'?

10. Why do we not grow bacteria at 37°C in schools or colleges?

---

students' book page 256

B3 3.1   Growing microbes

KEY POINTS

1 On a small scale, microbes are grown on agar in Petri dishes.
2 Carbohydrate is the energy source for the microbes.
3 Special precautions are taken to grow microorganisms safely and as uncontaminated cultures.
4 In schools and colleges cultures are incubated at 25°C.

Microorganisms are grown on agar in Petri dishes. Agar contains carbohydrate as an energy source, but may also contain some minerals, proteins and vitamins as supplementary (extra) nutrients.

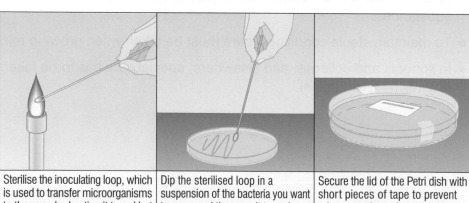

Sterilise the inoculating loop, which is used to transfer microorganisms to the agar, by heating it to red hot in the flame of a Bunsen and then leaving it to cool.	Dip the sterilised loop in a suspension of the bacteria you want to grow and then use it to make zig-zag streaks across the surface of the agar. Tilt the lid of the Petri dish to keep out unwanted microbes and close the lid as quickly as possible to avoid contamination.	Secure the lid of the Petri dish with short pieces of tape to prevent microorganisms from the air contaminating the culture – or microbes from the culture escaping. Do NOT seal all the way around the edge.

Culturing microorganisms in the lab

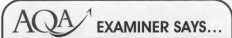
To prepare a pure or uncontaminated culture:

- Petri dishes and the culture medium (agar) must be sterilised before use.
- Inoculating loops, used to transfer the microorganisms, must be sterilised first by passing through a flame.
- The lid of the Petri dish should be held in place with adhesive tape to prevent any microorganisms getting in or out.

In schools and colleges these cultures should be incubated at 25°C. If grown at higher temperatures, especially 'body temperature', harmful microorganisms (pathogens) are more likely to grow. In industry higher temperatures are used to promote faster growth.

Key words: agar, culture, sterile, inoculating, incubate

CHECK YOURSELF

1 What is the energy source for growing microorganisms?

2 What name do we give to harmful microorganisms?

3 Which piece of equipment is used to transfer microorganisms to the Petri dish?

---

students' book page 258

B3 3.2 Food production using yeast

KEY POINTS

1 Yeast is a single-celled organism.
2 Yeast is used in making bread and alcoholic drinks.
3 Yeast respires aerobically and anaerobically.
4 The anaerobic respiration of yeast produces alcohol (and carbon dioxide).

Brewing at Ringwood Brewery in Hampshire. In tanks like these billions of yeast cells respire anaerobically, turning sugar into alcohol every day of the year.

Yeast cells have a nucleus, cytoplasm and a cell membrane surrounded by a cell wall. When oxygen is present, they respire aerobically and reproduce quickly.

If oxygen is absent, they respire anaerobically and produce alcohol (ethanol) and carbon dioxide. This is called fermentation.

When beer is brewed the starch in barley grains is used as the carbohydrate energy source for the yeast. The starch is first broken down into sugars in the barley grains by enzymes during the germinating process. When these sugars have been fermented, and the process is complete, hops are added to give the beer flavour.

When wine is made the grapes contain 'natural' sugars, which the yeast uses as the energy source.

Key words: aerobic, anaerobic, yeast, ferment, germinate, enzymes, hops

GET IT RIGHT!

For yeast to work well it has to respire aerobically first so that it can reproduce rapidly. If the conditions become anaerobic too quickly there will not be enough yeast cells to complete the process.

CHECK YOURSELF

1 Which cereal grains are used in beer making?

2 What is added to the beer to give it its taste?

3 Which chemicals break down the starches in the germinating barley grains?

B3 3.3 Food production using bacteria

1 Bacteria are used to help produce cheese and yoghurt.
2 In yoghurt production, bacteria convert lactose sugar into lactic acid.

In the UK we get most of our milk from cows, but around the world a number of different animals including camels, horses, sheep and goats are used for milking

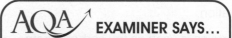 **EXAMINER SAYS...**

The most important part in the process of making yoghurt is the fermenting of lactose to produce lactic acid. The bacteria carry out this process

Bacteria are widely used in the manufacture of different cheeses. They are also used to make yoghurt.

There are three stages to making yoghurt:

- Bacteria are added to warmed milk.
- The milk sugar (lactose) is fermented by the bacteria, producing lactic acid.
- The lactic acid causes the milk to solidify (clot) and yoghurt is formed.

Key words: lactose, lactic acid, ferment, clot

Modern yoghurt production

In cheese-making, curds are formed by the action of one set of bacteria on the milk. Then the final flavour and texture of the cheese may well depend on other bacteria added at this stage of the process.

CHECK YOURSELF

1 What are the two types of food made using bacteria?

2 What is the name of the process that produces lactic acid?

3 What is the chemical name for milk sugar?

Large-scale microbe production

KEY POINTS

1 Microorganisms can be grown on a large scale in industrial fermenters.
2 The conditions in a fermenter must be very carefully controlled.
3 A mycoprotein can be produced using the fungus *Fusarium*.

AQA EXAMINER SAYS...

The organisms in a fermenter need a constant air supply and source of nutrients so that they are able to respire and produce energy. This energy releases heat, which is why the fermenter is constantly being cooled.

Industrial fermenters are large vessels used to grow microorganisms. The conditions in the fermenter are very carefully monitored:

- There is an air supply providing oxygen for respiration.
- There is a stirrer used to keep the microorganisms spread out and to make sure that the temperature is the same in all parts of the vessel.
- There is a water-cooled jacket around the outside, as the respiring microorganisms release heat and a constant temperature needs to be maintained.
- There are sensors to monitor both pH and temperature.

The fungus *Fusarium* is grown to produce mycoprotein, a protein-rich food suitable for vegetarians. *Fusarium* is grown aerobically on starch and the mycoprotein harvested.

Key words: sensors, *Fusarium*, mycoprotein

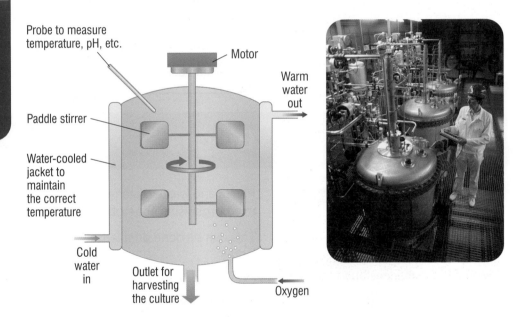

Probe to measure temperature, pH, etc.
Motor
Warm water out
Paddle stirrer
Water-cooled jacket to maintain the correct temperature
Cold water in
Outlet for harvesting the culture
Oxygen

The design of fermenters is improving all the time as new ways are developed of keeping conditions inside the fermenter as stable as possible

CHECK YOURSELF

1 Name two conditions in a fermenter that are monitored closely.

2 Why does a fermenter need to be kept cool?

3 Suggest why mycoprotein is valuable to grow.

B3 3.5 Antibiotic production

EXAMINER SAYS…

Penicillium does produce penicillin, but only when most of the nutrients in the fermenter have been used up.

Penicillin is made from the mould *Penicillium* in a fermenter.

The medium (solution inside the vessel) contains sugar for energy and some other nutrients including nitrogen. *Penicillium* only starts to produce the antibiotic when most of the nutrients are used up.

Key words: penicillin, antibiotic, *Penicillium*

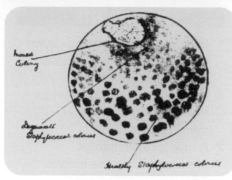

Alexander Fleming noticed the clear areas on his plates. He realised he had made a discovery of enormous potential.

CHECK YOURSELF

1 What is the name of the mould used to produce penicillin?

2 Why is sugar used in the fermenter?

3 Which other important nutrient is used by the mould?

B3 3.6 Biogas

GET IT RIGHT!

The gas is produced through fermentation. The microorganisms respire anaerobically to produce the gas.

Plants and waste material (containing carbohydrate) can be broken down by microorganisms anaerobically to produce biogas. The gas is mainly methane.

This can be done on a large scale with waste from sugar factories or sewage works.

On a small scale it can be used by a home or farm. The gas produced is a fuel and provides energy.

Key words: methane, fuel, energy

Biogas generators like this have made an enormous difference to many families by producing cheap and readily available fuel

CHECK YOURSELF

1 What is the main gas in biogas?

2 What is the energy source for the microorganisms in the process producing the gas?

3 What are the microorganisms doing to produce biogas?

B3 3.7 More biofuels

1 Ethanol-based fuels can be produced by fermentation and used to power cars.
2 Microorganisms respire anaerobically to produce the ethanol, using sugars as the energy source.
3 The ethanol produced through the fermentation process has to be distilled before it can be used.

The Sun and the rain in areas like the Caribbean allow plants like this sugar cane to photosynthesise and grow very rapidly. The next step is to convert them into useable fuel.

Sugar cane juices and glucose, derived from maize starch by the action of a carbohydrase enzyme, can be fermented to produce ethanol. Microorganisms respire anaerobically in this process.

The ethanol produced must be distilled. It can then be used as a fuel in motor vehicles.

Key words: ethanol, distilled, carbohydrase enzyme

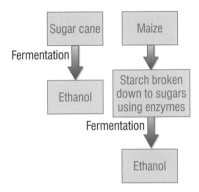

EXAMINER SAYS...

Using ethanol as a fuel could replace fossil fuels in the future. In terms of the 'greenhouse effect', using ethanol as a fuel is much more 'carbon friendly'. You could be asked in an examination for the advantages and disadvantages of using ethanol.

The starch in maize needs to be broken down by enzymes before yeast can use it as a fuel for anaerobic respiration. Although it takes more steps to produce ethanol from maize than from sugar cane, maize can be grown in many more countries around the world.

CHECK YOURSELF

1 What types of enzymes digest the maize starch to produce glucose?

2 How do the microorganisms respire during the fermentation process?

3 What must be done with the ethanol the microorganisms produce before the ethanol can be used as a fuel?

B3 3 End of chapter questions

1 **What type of dish are microorganisms grown in?**

2 **How are inoculating loops sterilised?**

3 **Why is the lid sealed onto the Petri dish after the culture has been set up?**

4 **What are the waste products of the anaerobic respiration of yeast?**

5 **Why is yeast first allowed to respire aerobically during the production of beer?**

6 **Why is it important to keep the pH constant during growth in a fermenter?**

7 **What is the energy source for the growth of *Fusarium*?**

8 **At what stage in the fermenter does the mould *Penicillium* start to produce penicillin?**

...teria are used to make yoghurt.

(a) The process is started by adding a culture of bacteria to milk. Why should the milk be warmed first?
(2 marks)

(b) (i) What substance in the milk is fermented?
(ii) What is the product of this fermentation?
(iii) Why does the milk solidify to produce yoghurt?
(3 marks)

2 Microorganisms can be grown in vessels called fermenters. This is a diagram of a fermenter.

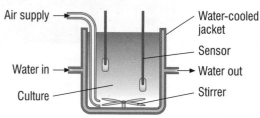

(a) (i) Why does the fermenter have a good air supply?
(2 marks)
(ii) The fermenter is cooled by a water-filled jacket. What is causing the temperature inside to increase?
(2 marks)
(iii) State one other condition in the fermenter that needs to be monitored and controlled. (1 mark)

(b) The antibiotic penicillin is produced in fermenters.
(i) What is the name of the mould that produces penicillin?
(1 mark)
(ii) What nutrients does the growth medium contain?
(2 marks)
(iii) At what point in the process does the mould actually start producing penicillin? (1 mark)

(c) It is thought that many infections are transferred in hospitals from patient to patient by doctors and nurses. Semmelweiss was a Hungarian doctor who worked in Vienna. He carried out an experiment with broth in glass flasks. He boiled the broth to kill all of the organisms in the flasks and then left some flasks open but sealed the others shut.
(i) What was he trying to prove? (1 mark)
(ii) After his work, what action did he suggest that medical staff take in the hospitals? (1 mark)
(iii) Why did he suggest they do this? (2 marks)
(iv) Why were his ideas not accepted until some time later? (2 marks)

3 Yeast is used in the brewing of beer.

(a) What do the brewers do with the barley grains so that they provide an energy source for the yeast? (2 marks)
(b) (i) Why is the yeast allowed to respire aerobically at the start of the process? (1 mark)
(ii) What is the name of the process that results in the production of alcohol by the yeast? (1 mark)
(iii) What is the name of the alcohol produced? (1 mark)
(iv) What is the name of the other waste product in this process? (1 mark)
(c) What is added to the beer to give it flavour? (1 mark)
(d) Yeast is also used to make wine from grapes. What is in the grapes that provides an energy source for the yeast? (1 mark)

4 As fossil fuels begin to run out other sources of energy are being developed. Biogas generators are now quite widely used. The diagram shows a simple biogas generator.

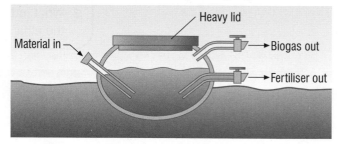

(a) Which is the main gas produced by these generators? (1 mark)
(b) (i) What is put in these generators that is used by the bacteria to produce the gas? (2 marks)
(ii) What process in the bacteria cells breaks down the material? (2 marks)
(c) Suggest why small biogas fermenters might be particularly useful for families living in remote areas of developing countries (2 marks)
(d) In which parts of the United Kingdom might you expect to find biogas generators that are large and produce a lot of energy? (2 marks)

 Test & Assessment Interactive quizzes, answers and hints online!

The answer is worth 3 marks out of the 5 marks available.

The responses worth a mark are underlined in red.

We can improve the answer in several ways:

Oil and petrol are fossil fuels and at some stage in the future will run out. Burning them also produces carbon dioxide as well as nitrogen oxides. In some countries, for example Brazil, sugar cane is used to produce ethanol. The ethanol is mixed with petrol and used as a fuel for cars. This fuel produces carbon dioxide but much less nitrogen oxide when burned. Rainforests have been cut down to create more space to grow the sugar cane.

It is said that this new fuel is better for the environment and is 'carbon neutral'. Use information from the passage and your own knowledge to state and explain your view on whether this new fuel is better or worse for the environment.

(5 marks)

The ideas stated about the Greenhouse effect are good but there is no worthwhile mention of the idea that this new fuel is 'carbon neutral'.

It is better for the environment as fossil fuels are running out and we need something else. It also produces less carbon dioxide and so there is not as much Greenhouse effect which is causing the world to warm up and the ice caps to melt. It could be carbon neutral as when it burns it gives off as much carbon dioxide as it uses.

This is too vague. The plant **takes up CO$_2$ (during photosynthesis)** and then releases it when the fuel burns is the key point. However the fact that rainforests are cleared to grow the sugar cane **reduces the amount of CO$_2$ taken up,** and is bad for the environment in terms of **habitat destruction**. Less nitrogen oxides would mean **less acid rain** – another valid point in favour of the new fuel.

The answer is worth 2 marks out of the 4 marks available.

The responses worth a mark are underlined.

We can improve the answer in the following ways:

You are given a Petri dish containing agar and a small bottle with a culture of bacteria growing inside. How would you grow a pure culture of the bacteria in the Petri dish?

(4 marks)

You would use a wire loop to get some bacteria and put them on the agar. You would put the lid on and sellotape it down to stop more bacteria getting in. You would do it all keeping everything sterile. You would keep it in a warm place.

You should sterilise the (inoculating) loop by passing it **through a flame** before collecting the bacteria.

You should state that you would **incubate your culture at about 25°C**, and not at 37°C as pathogens might grow at this temperature.

This is not precise enough. As well as passing the loop through a flame, you would also **flame the neck of the bottle** the bacteria were in.

Chapter 1

Pre Test

1 Electrical along neurones, chemical across a synapse.
2 In glands.
3 The gap between two neurones.
4 A receptor.
5 28 days (in most women).
6 Follicle stimulating hormone (FSH), oestrogen, luteinising hormone (LH).
7 Luteinising hormone.
8 Follicle stimulating hormone and luteinising hormone.
9 Oestrogen.
10 Follicle stimulating hormone.

Check yourself

1.1
1 Transported by the bloodstream.
2 A gland.
3 Electrical in neurones, chemical across a synapse.

1.2
1 By chemicals.
2 To pass the impulse on to the correct motor neurone.
3 Glands or muscles that make a response.

1.3
1 In the pituitary gland.
2 Oestrogen inhibits the production of FSH and stimulates the production of LH.
3 LH stimulates the release of the mature egg from the ovary.

1.4
1 Oestrogen.
2 FSH stimulates eggs to mature in the ovaries.

1.5
1 We sweat, produce urine and it is in our breath when we breathe out.
2 Our enzymes only work effectively within a narrow temperature range.
3 It is the energy source for respiring cells.

End of chapter questions

1 A stimulus is detected by a receptor. A sensory neurone is stimulated and carries the impulse to the relay neurone. The relay neurone passes the impulse to the correct motor neurone.
The motor neurone carries the impulse to an effector (gland or muscle).
The effector makes a response.
Between the neurones the impulse crosses gaps, called synapses, by means of chemicals.
2 The hormone system relies only on messages being transmitted by chemicals. Impulses in the nervous system are electrical, except at synapses where they are chemical. (The hormone system also tends to be slower with the effects being longer lasting.)

3 FSH stimulates both eggs to mature and the release of oestrogen. Oestrogen inhibits the further production of FSH, stimulates the lining of the womb to build up ready for pregnancy and stimulates the pituitary gland to make LH.
LH stimulates mature eggs to be released.
4 An argument for would be that it allows women (couples) to have children when they would otherwise not be able to. Arguments against might include the cost, it is not 'natural', it is against 'God's will', many eggs may be fertilised but most will 'die', there have been multiple births with most embryos not surviving.
5 Two of: water content, ion content, temperature, blood sugar level.
6 Relay.
7 It contains oestrogen, which inhibits the production of FSH.
8 In the pituitary gland.

Chapter 2

Pre Test

1 Eating the right amounts of the different foods that you need.
2 Not eating the right balance of these different foods (it doesn't just mean not eating *enough* food).
3 The sex they are, how active they are, whether they are pregnant, the temperature they are living in.
4 The rate (speed) that energy is being released from food by the living cells.
5 Obese.
6 Arthritis, diabetes, high blood pressure, heart disease.
7 In the liver.
8 High density and low density (HDLs and LDLs).
9 Saturated fat.
10 High blood pressure.

Check yourself

2.1
1 The rate at which energy is released from food by living cells.
2 Two of: if it is colder, if you are more active, if you are pregnant.
3 You need more energy for the activity. The cells respond by releasing energy from food more quickly.

2.2
1 Obese.
2 A disease of the joints.
3 High blood pressure, diabetes, heart disease.

2.3
1 High blood pressure.
2 They transport cholesterol.
3 Mono-unsaturated and polyunsaturated fats.

End of chapter questions

1 Statins.
2 Saturated fat.
3 Three of: arthritis, diabetes, high blood pressure, heart disease.
4 The rate at which your living cells are releasing energy from food.
5 To replace the heat that you are losing.
6 Reduced resistance to infection or irregular periods for women.
7 Low density lipoproteins.
8 Salt and/or fat.

Chapter 3

Pre Test

1 Against the law.
2 To make very sure that there are no side-effects.
3 As a sleeping pill, but it also helped to prevent 'morning sickness' during pregnancy.
4 The effects of stopping taking a drug on the body.
5 Drugs that people take to stimulate them simply because they want to. They are not being taken for a medical reason.
6 Two of: cannabis, heroin, cocaine.
7 Heroin, cocaine.
8 Nicotine.
9 It is taken up by red blood cells instead of oxygen. You will have less energy or your heart and lungs will have to work harder to compensate.
10 It slows down your reactions.

Check yourself

3.1
1 Leprosy.
2 One of: cannabis, heroin, cocaine, alcohol, tobacco (nicotine).
3 Many more people take them.

3.2
1 Alcohol, tobacco (smoking).
2 Pressure from the people you are with, wanting to experiment further.
3 Many more people take legal drugs.

3.3
1 It slows down your reactions.
2 The brain and liver.
3 Your reactions are slowed so you take longer to stop. You lose self-control, so you may drive more dangerously or may not be able to judge what is happening on the road very well. You may go into a coma if you have drunk a great deal of alcohol.

3.4
1 There are cancer causing chemicals (carcinogens) in the smoke.
2 Nicotine.
3 They do not get enough oxygen so cannot release as much energy from the food they receive as they should. The carbon monoxide in the smoke reduces the oxygen carrying capacity of the baby's blood.

End of chapter questions

1 It slows it down. Impulses travel more slowly.
2 It reduces the oxygen carrying capacity of the blood. Less energy is released from food. Heart and breathing rate may go up to try to compensate for this.
3 You react more slowly. You may not care as much (lack of self-control). Your judgement will not be as good – both in terms of how you drive and whether you are likely to cause an accident.
4 They are in the tobacco smoke.
5 They need to be tested to see whether they are toxic or have any side effects. Side effects sometimes take many years to be seen.
6 Once a person has tried cannabis he/she may want to try something else, perhaps something more dangerous.
7 Carcinogen.
8 Tobacco or alcohol.

Chapter 4

Pre Test

1 An organism that causes an infectious disease.
2 Bacteria and viruses.
3 Toxins.
4 White blood cell.
5 A chemical that neutralises (cancels out) a toxin (poison).
6 Relieve the symptoms of the disease but they *do not* cure it.
7 A chemical, produced by white blood cells, which helps to destroy pathogens.
8 They live inside the cells. To destroy a virus, you often end up destroying the cell as well.
9 It changes and is resistant to most commonly used antibiotics.
10 A vaccine is made from dead or inactive forms of the pathogen. It is used to give immunity to the person.

Check yourself
4.1

1 Bacteria and viruses.
2 By stopping the transfer of infection from one patient to another.
3 Doctors could not see what was supposed to be causing the infection, they thought it was God's punishment to women, they did not want to admit that they might be responsible for the deaths of patients and hand washing was a strange idea at the time anyway!

4.2

1 By stopping the pathogens entering the body in the first place.
2 They can ingest pathogens and they produce antitoxins, which neutralise the toxins that pathogens produce.
3 They neutralise the toxins.

4.3

1 They kill bacteria but find it more difficult to kill viruses.

2 Viruses live and reproduce inside the body's cells.
3 To alleviate (reduce) the symptoms of the infection (*note:* they do not help to cure the infection).

4.4

1 A disease that spreads across a number of countries.
2 MRSA is resistant to the antibiotics that we normally use.
3 The best-adapted members of a species survive and breed. They pass on their genes. 'Weaker' members of the species do not survive and do not pass on their genes through reproduction.

4.5

1 The drugs have to be tested on volunteers. It can take years for side effects to develop. Until the company is sure there are no side effects, it will not sell the drug. It will cost them a lot of money if it proves to be dangerous.
2 As a sleeping pill and to prevent morning sickness.
3 They believe it is cruel. It causes the animals pain and suffering.

4.6

1 To be injected with a vaccine, which contains dead or inactive forms of the pathogen.
2 The body produces antibodies in response to the vaccine. If the body is infected with the pathogen, it responds very quickly in making antibodies. The disease is destroyed before you even begin to feel ill.
3 If they were not, you might be infected with the disease.

End of chapter questions

1 That infection is spread from one person to another.
2 Pathogens produce toxins (poisons), which make you feel ill.
3 A vaccine contains dead or inactive forms of the pathogen.
4 Antibodies are produced by white blood cells. They help to destroy pathogens in the blood.
5 It means to digest or to destroy them.
6 Penicillin.
7 In the living cells.
8 A change in an organism.

EXAMINATION-STYLE QUESTIONS

1 (a) (i) Eyes.
    (ii) Ears.
    (iii) Skin (surface).
    (iv) Ears. (4 marks)
(b) The nervous system works through electrical impulse; the hormone system is chemical.
The nervous system tends to be faster, but the hormone system tends to have a longer lasting effect. (2 marks)
(c) There are many, any two from: fast, strong, large teeth, large claws, camouflage, good sight.
(2 marks)

2 (a) A drug used for pleasure / not a medicine. (1 mark)
(b) Two of: heroin, cocaine, cannabis. (2 marks)
(c) (i) Brain and liver. (2 marks)
    (ii) Slows it down. (1 mark)
(d) (i) Nicotine. (1 mark)
    (ii) Two of: lung cancer, emphysema, heart disease. (2 marks)
    (iii) the physical pain / longing when not taking the drug (1 mark)
    (iv) Smoke contains carbon monoxide, this gets into the baby; it combines with the red blood cells so less oxygen is carried; the baby is able to release less energy from food; the baby doesn't grow as much. (4 marks)

3 (a) They produce toxins (poisons). (1 mark)
(b) The vaccine contains dead (or inactive) forms of the pathogen; the white blood cells respond by producing antibodies; antibodies are made very quickly if the pathogen enters the body again. (3 marks)
(c) Viruses live inside the cells. If you kill the virus you are likely to damage the cell. (2 marks)
(d) (i) That infection was transferred between patients in hospitals. (2 marks)
    (ii) Doctors did not want to admit that they were responsible; hand washing was thought to be a strange idea, it was thought to be punishment for the women. (3 marks)

4 (a) The correct amount of all the different foods. (1 mark)
(b) Three of: sex, size, temperature, exercise, pregnancy. (3 marks)
(c) (i) It can give you high blood pressure. (1 mark)
    (ii) Three of: diabetes, high blood pressure, arthritis, heart disease. (3 marks)
(d) (i) In the liver. (1 mark)
    (ii) Disease of the blood vessels; disease of the heart (2 marks)
    (iii) Any three: by eating less saturated fat; by eating proportionally more polyunsaturated fat; by increasing cholesterol transport by low density lipoproteins (LDLs); by taking statins. (3 marks)

Chapter 5

Pre Test
1 Light, water, nutrients.
2 One of: mates, territory.
3 So that it is camouflaged against the snow and ice. It is harder for its prey to see it.
4 The snow and ice has melted. The earth is brown. The prey is camouflaged, so that it is more difficult for its predators to see it.
5 How big the surface area of its skin is compared to the whole size of its body. (This is why a mouse has a large surface area compared to the size of its body, but an elephant has a small surface area compared to the volume of its body.)
6 The heat or the lack of water (and with animals, therefore, the possible lack of food to eat).
7 So that those that eat them know that they have a horrible taste or are poisonous. Once they have tasted them they won't do it again! With animals bright colours are also used to attract a mate.
8 So that they can find enough food (and water). They also need space to breed in.
9 Light, water, nutrients.
10 So that the new plants don't grow alongside them and compete.

Check yourself
5.1
1 It will find it difficult to lose heat, however it will also not heat up as fast as smaller animals.
2 They lose heat through their relatively large surface area quickly and may not be able to find enough food to generate the energy they need.
3 So they are camouflaged against the snow and ice and their prey find it more difficult to see them.

5.2
1 Anywhere where taller plants, e.g. trees, grow. This might be in a forest or woodland.
2 Three of: smaller leaves, water storage in the stem, holes in the leaves (stomata) out of the light and any wind, waxy leaves.
3 The plants may be poisonous or have an awful taste. They have bright colours warning the animals off.

5.3
1 So that they can find enough food (and water). They also need space for breeding.
2 They are showing that they are poisonous or have an awful taste. Once an animal eats it, they will remember the experience!
3 They have long necks, so are able to get to food that other animals cannot

reach. This is particularly important when food is in short supply.

5.4
1 Light, nutrients and water.
2 So that they are able to use the sunlight to grow. Once the trees have their leaves the smaller plants will receive much less light. They will also have much more competition for nutrients and water as the trees begin to grow.
3 If they grow beside the parent plant they will compete with the parent for resources.

End of chapter questions
1 If it eats a plant it doesn't like, and that plant species has a warning colour, it is more likely not to eat that plant again.
2 For mates, food, territory.
3 So that their prey find them more difficult to see. This means that the predator can get closer to its prey before attacking, so there is more chance of success.
4 The heat and the lack of water.
5 To keep them warm in cold seas.
6 One of: very small leaves (spines), stems that photosynthesise, storage of water in the stem.
7 To keep cool / out of the Sun's rays, to catch prey as they cannot be seen to avoid being eaten.
8 They find out which is the slowest, so saving themselves some work.

Chapter 6

Pre Test
1 Male and female sex cells.
2 In the nucleus.
3 Genes (or DNA).
4 Genes from the male and female parent are mixed.
5 It does not lead to any variation. The offspring are genetically identical to the parent.
6 They contain exactly the same genes as the parent. There is no variation (asexual reproduction).
7 You take a group of cells from the plant. You grow them in a special medium, where they develop roots and shoots, before growing them on in soil or compost.
8 Producing identical organisms to the parent through asexual reproduction.
9 Using enzymes.
10 Those against may think it is messing about with nature or that it is in conflict with God's work. Some against feel that the genes that are moved from organism to organism may find their way into other organisms you don't want them to be in, so control is lost. Those in favour feel that it will result in improved organisms with more of the 'desired' characteristics that we want.

Check yourself
6.1
1 The characteristics of an organism.
2 The information is passed on by the sex cells (gametes) in the genes.
3 In the nucleus.

6.2
1 Sexual reproduction.
2 The genes from both parents are mixed together when the sex cells fuse together.
3 Clones.

6.3
1 Tissue culture and cuttings.
2 The embryo has a certain genetic make up from the parents' sex cells. If you just split the cells apart you are not mixing any new genes into the cells, so they will remain the same as each other and the original embryo.
3 The animal in which the embryo develops.

6.4
1 An adult udder cell.
2 An egg cell.
3 A mild electric shock is applied.

6.5
1 Enzymes are used to cut the gene out.
2 It will develop characteristics associated with the new gene.
3 So that organisms can be developed with the 'desired' characteristics we want, e.g. plants or animals that produce more food or are resistant to disease, etc.

End of chapter questions
1 So that we can produce lots of plants that we want quickly and cheaply and we know that they will be the same as the original plant.
2 The male and female sex cells each have their complement of genes from each parent. When the cells fuse together these genes mix so the offspring is similar to, but not the same as, either parent.
3 Embryo transplants, adult cell cloning, fusion cell.
4 Taking a few cells from a plant and growing them in a special medium so that they develop roots and shoots. The new plant is then grown in the normal way in compost or soil.
5 DNA/genes/chromosomes.
6 There is no mixing of genes / the offspring have the same genetic make-up as the parent.
7 Cuttings, as it is much cheaper and much quicker.
8 Insulin.

Chapter 7

Pre Test
1 All organisms in the same species

vary. Some will be more successful and breed. Their genes are passed on to the next generation. Unsuccessful members of the species do not pass their genes on.
2 That if an organism acquired a characteristic, e.g. if a human, through training, developed a fast speed when running, this characteristic would be passed on to the offspring.
3 People can see that we do not pass on these 'acquired characteristics' to the next generation.
4 Many people believed in God and thought that God created new species so the church was very much against Darwin's ideas.
5 They contain the remains of organisms that lived a long time ago. These fossils show that organisms of a species changed only slowly over very long periods of time. In Lamarck's theory, there would be very big changes over very short periods of time.
6 About 3500 million years ago.
7 The ones that are well adapted to their environment and successfully find a mate.
8 Through the sex cells (gametes).
9 When all members of a species 'die out'.
10 A change in a gene.

Check yourself
7.1
1 3500 million years ago.
2 No-one was here and there are two conflicting ideas. Both of these ideas are possible and no-one can either prove one idea or disprove the other one!
3 Through the rocks that they are found in. We can date these rocks.
7.2
1 One which the organism develops while it is alive.
2 Many die out through lack of food, disease, being eaten, being too hot or too cold.
3 They preferred to believe that it was God that created new species.
7.3
1 If new members of the species are produced by sexual reproduction, then the genes from both parents are mixed together when the sex cells fuse.
2 Those that are best adapted or suited to their environment survive and live long enough to breed. The weaker ones do not survive.
3 The organisms best adapted to their environment survive and breed, passing their characteristics on to the next generation.
7.4
1 All of the organisms of the species die out.
2 It may eat all of the food of that species.
3 By hunting them for food, by removing the habitat of the species, through polluting the environment, by introducing new competitors to the species.

End of chapter questions
1 That if an organism acquired a characteristic during its lifetime it could pass this on to its offspring.
2 That simple organisms arrived here from another planet or that conditions on the Earth were such that, along with the energy from lightning, life-forms developed.
3 Two of: competition, disease, new predators, change in the environment, destruction of habitat.
4 We can date rocks. Fossils of previous life-forms are found in rocks.
5 4500 million years.
6 Those that are best adapted to their environment survive and are able to breed.
7 It could fly, swim or be introduced by humans (accidentally or deliberately).
8 Temperature, rainfall.

Chapter 8
Pre Test
1 Once these resources have run out there will be no more as they take so long to be created.
2 Two of: farming, building, quarrying and dumping waste.
3 With sewage, fertiliser, toxic chemicals directly into the water. Pesticides and herbicides can also get washed off the land into water.
4 A substance that enables plants to grow better / replaces nutrients in the soil.
5 Sulfur dioxide.
6 A chemical that kills the pests that can affect plant growth.
7 Invertebrate animals.
8 Carbon dioxide and methane.
9 The trees take in the carbon dioxide for photosynthesis.
10 Development that does not reduce the resources available on the Earth, including land.

Check yourself
8.1
1 Building, quarrying, farming, dumping waste.
2 Pesticides and herbicides.
3 Non-renewable.
8.2
1 Sulfur dioxide and nitrogen oxides.
2 Clouds are blown by the wind so the rain may fall on another country.
3 Enzymes work best in a very narrow range of acid / alkali conditions (pH). The acidity of the rain stops them working and may denature them.
8.3
1 By cows and by rice fields.
2 Much of the heat of the Sun is radiated back out in to space. Some is absorbed by the atmosphere. The more greenhouse gases there are, the less heat is radiated back into space. The Earth heats up.
3 If trees are cut down, they will not take carbon dioxide up for photosynthesis.

If the trees are burned or are left to decompose, then they will release the carbon they contain as carbon dioxide.
8.4
1 The fossil fuels we are using, e.g. petrol, diesel and kerosene, are all non-renewable and will run out.
2 There is only a limited area of land on the Earth. If we keep using more and more, there is less left for other animals and plants.
3 Aluminium is a limited resource. It will eventually run out. It also takes a lot of energy to extract aluminium from its ore.
8.5
1 So as not to destroy too much habitat in the countryside.
2 If they are living in the water you are testing you immediately know it is polluted.
3 Wider.

End of chapter questions
1 Two of: building, farming, quarrying, dumping waste.
2 Lichens.
3 Enzymes.
4 By decomposing (the respiration of the bacteria) or by being burned (combustion).
6 Two of: sewage, fertiliser, toxic chemicals.
7 Carbon dioxide.
8 To kill weeds.
9 Two of: paper, aluminium (cans), plastic, glass.

EXAMINATION-STYLE QUESTIONS
1 Gametes, genes, chromosomes, nucleus.   (4 marks)
2 (a) Two of: building, dumping waste, quarrying, farming.   (2 marks)
   (b) (i) To help plants grow/replace soil nutrients.
       (ii) To kill weeds in the crops.
       (iii) To kill unwanted pests (e.g. insects).   (3 marks)
   (c) The chemicals get onto the soil, it rains and they are washed into ponds / rivers (leaching). (2 marks)
   (d) A resource that cannot be replaced – it takes too long to create.
                                              (1 mark)
3 (a) (i) The genes / DNA carries the genetic information; the genes / DNA is from the pet; the new embryo develops into the new pet / cell division; there is no mixing of genes / asexual reproduction.   (4 marks)
       (ii) Adult cell cloning.   (1 mark)
   (b) Two of: It is expensive; it is against nature / God; many embryos die for each one that is successful; the new animal is more likely to have health problems.   (2 marks)
   (c) Cuttings – take a piece of the plant and grow it on (in compost). Tissue culture – take a small group

of cells and grow further in a special growth medium. (4 marks)

4 (a) (i) The length of the giraffes' necks shows variation; this is due to sexual reproduction / the mixing of genes; when food is in short supply, giraffes with the longer necks are more likely to survive (natural selection), this is 'survival of the fittest'; these giraffes breed and pass on their genes. (4 marks)
   (ii) To get to food, giraffes would have 'stretched' their necks, they would have passed this 'acquired characteristic' on to their young. (2 marks)
(b) (i) 3500 million years ago. (1 mark)
    (ii) It arrived from another planet, possibly by meteorites, *or* there was a particular mix of chemicals on Earth; lightning provided the energy to create life from this mix. (4 marks)
(c) Fossils can be found in rocks. It is possible to date rocks. (2 marks)
(d) Two of: Competition – new organisms arrived and competed successfully for food.
Disease – a new disease evolved and the organisms had no immunity to it.
Change in conditions – weather conditions changed, e.g. too warm / cold, the organisms could not adapt quickly enough.
Predation – a new predator came along and effectively killed all of the species for food. (2 marks)

B2 Answers to questions

Chapter 1

Pre Test
1 It controls the activities of the cell.
2 The cell membrane.
3 The cytoplasm.
4 To produce proteins (protein synthesis).
5 Chloroplasts (containing chlorophyll).
6 Mitochondria.
7 There are many different functions (jobs) that plant and animal cells have to carry out.
8 The random movement of particles from an area of high concentration to an area of lower concentration.
9 Many examples, e.g. oxygen, carbon dioxide, glucose.
10 It takes place through a partially permeable membrane and involves only the movement of water.

Check yourself
1.1
1 Three of: cell membrane, nucleus, cytoplasm, mitochondria, ribosomes.
2 In the mitochondria.
3 A number of examples including carbon dioxide, oxygen and glucose.
1.2
1 Egg cells, muscle cells, white blood cells (many others).
2 It is long and has many endings to communicate with other nerves or effectors.
3 Differentiation.
1.3
1 They naturally vibrate as they have energy.
2 The diffusion of glucose into respiring cells.
3 Cells need a constant supply of materials and to get rid of waste products.
1.4
1 It involves the random movement of particles.
2 A membrane that only allows the smaller molecules (particles) through.
3 For support, and chemical reactions take place in solution.

End of chapter questions
1 To synthesise (make) proteins.
2 It is streamlined for swimming and has lots of mitochondria to produce energy.
3 Particles naturally vibrate – the cell does not need to make them move.
4 It also passes from the stronger to the weaker solution, as the movement of the water molecules is random. However the net movement is to the stronger solution.
5 Chlorophyll.
6 It is for support.
7 It is at a higher concentration in the blood than in the cell.
8 The cell membrane.

Chapter 2

Pre Test
1 The Sun.
2 Glucose.
3 Carbon dioxide.
4 The enzymes do not work so effectively and the molecules move more slowly.
5 There is more light so more energy for the process, and more heat so the molecules move faster colliding more often and with more energy.
6 The temperature will limit the rate of photosynthesis anyway, so increasing the amount of light would have no effect.
7 In dense vegetation, where photosynthesis is taking place rapidly and temperature is high, e.g. in a tropical rainforest.
8 To make their own proteins for growth.
9 Stunted.
10 They cannot make green chlorophyll.

Check yourself
2.1
1 The Sun.
2 Carbon dioxide and water.
3 Oxygen is needed by plants and animals for respiration – the release of energy from food.
2.2
1 The molecules move more slowly, so collisions are less forceful and occur less often.
2 It will be hot enough anyway so you will be wasting energy. It may even become so hot that enzymes denature.
3 In a rapidly photosynthesising area where it is warm, e.g. in a tropical rainforest.
2.3
1 So that the substance does not affect osmosis (as it does not dissolve and alter the strength of a solution).
2 It is combined with other nutrients to form new materials.
3 The release of energy from food.
2.4
1 Through the roots.
2 To make chlorophyll for photosynthesis.
3 It means a 'lack' of something.

End of chapter questions
1 It is trapped by chlorophyll.
2 Temperature.
3 So that it has no effect on osmosis.
4 It will be stunted (shorter).
5 Glucose (sugar).
6 At the bottom of hedgerows, smaller plants in a field of tall growing crops, behind a building, behind a wall – anywhere there is some shade.
7 They cannot make enough protein, protein is necessary for growth.
8 It is stored as starch.

Chapter 3

Pre Test
1 A picture showing the mass of living material at each stage of a food chain.
2 One organism, e.g. a bush, can support large numbers of other organisms so the pyramid can have a very unusual shape. However the mass of the tree is much greater than the mass of the organisms it supports.
3 A number of reasons including: not all the food is digested, the animal needs to keep warm, the animal will move around using up energy.
4 Birds need to maintain a constant temperature. The colder it is, the more energy they need to keep warm.

5 It is expensive as a lot of energy is lost when animals eat plants – not all of the plant is converted into meat (perhaps 20%).
6 By keeping the number of stages in the food chain as few as possible.
7 Decay or decomposition.
8 Microorganisms (bacteria and fungi).
9 Photosynthesis.
10 All of the nutrients are recycled and are re-used by organisms.

Check yourself

3.1
1 The mass of living material.
2 A tree can support many caterpillars, which support a few birds.
3 Each stage can be drawn in proportion to the other stages depending on the mass of organisms.

3.2
1 The energy at each stage that is not converted into energy in the next stage.
2 Because to keep warm, as you lose heat to the surroundings, requires a lot of energy.
3 Energy is required by muscles to contract and cause movement.

3.3
1 Less energy wastage between the stages of the food chain.
2 It will lose a lot of heat to the surroundings and will need energy to replace the energy lost.
3 They prevent the animal from any movement so that it uses less energy.

3.4
1 An organism that feeds on waste.
2 Enzymes work faster in warmer conditions.
3 Decomposition or break down of material.
4 Bacteria or fungi.

3.5
1 Respiration.
2 Detritus feeders or decay organisms (microorganisms).
3 Returning it to the atmosphere so that it is not all 'locked' up by organisms.

End of chapter questions
1 Respiration.
2 It costs too much in energy losses from plants to produce it. There would not be enough to eat.
3 The enzymes work more quickly.
4 Plants remove carbon from the atmosphere in photosynthesis. Plants return carbon to the atmosphere in respiration.
5 Biomass shows the amount of living material at each stage. With a pyramid one tree may support several thousand insects, so the pyramid does not accurately reflect what is happening.
6 It passes out of an animal as faeces and enters the decay process.
7 Detritus feeders.
8 They respire 24 hours per day, all of the time.

EXAMINATION-STYLE QUESTIONS

1 (a) A: cell membrane.
   B: nucleus.
   C: cytoplasm. (3 marks)
  (b) Long – impulses travel long distances.
   Lots of connections – to other nerves. (2 marks)
  (c) (i) Any two of: vacuole, chloroplast, cell wall. (2 marks)
   (ii) A vacuole contains the cell sap.
    A chloroplast contains chlorophyll / traps energy.
    A cell wall – adds strength to the cell. (2 marks)

2 (a) From the top down:
   Large bird (sparrowhawk), small bird (blue tit), caterpillar, tree. (3 marks)
  (b) Any three of: Some is not digested.
   Some is excreted.
   Some is used to keep warm.
   Some is used for movement. (3 marks)
  (c) Keep the animals at a warmer temperature.
   Stop the animals from moving around. (2 marks)

3 (a) (i) Respiration / energy. (1 mark)
    All of the time / 24 hours per day. (1 mark)
   (ii) To produce amino acids or to produce proteins.
    For growth. (2 marks)
   (iii) Small / stunted. (1 mark)
  (b) The leaves are yellow, as they cannot produce chlorophyll. (2 marks)
  (c) Warm and wet. (2 marks)
  (d) All of the materials are recycled within the community. (1 mark)

4 (a) (i) The (random) movement of particles, from an area of high concentration, to an area of lower concentration. (3 marks)
   (ii) A: carbon dioxide.
    B: oxygen. (2 marks)
  (b) The particles are moving (randomly) anyway.
   The energy source is the warmth / temperature / Sun. (2 marks)
  (c) (i) Any two of: Movement of particles (random), from higher to lower concentration, does not need energy from the cell. (2 marks)
   (ii) It only involves water molecules.
    It takes place through a partially permeable membrane. (2 marks)

Chapter 4

Pre Test
1 A catalyst that works in a living organism – an enzyme.
2 Respiration and photosynthesis (though there are many more).
3 It is where the reacting molecules (particles) fit.

4 In the mitochondria.
5 Proteins.
6 A gland.
7 Lipases.
8 In the liver.
9 In the stomach.
10 It is sweeter than glucose so you need less to sweeten foods. It is therefore useful in 'slimming' foods.

Check yourself

4.1
1 Proteins.
2 The area of the enzyme that the reacting particles fit into.
3 Speeding up a chemical reaction.

4.2
1 Enzymes work more effectively, molecules move faster so have more collisions and collide with more force.
2 Their shape changes, they denature. The reacting molecules will not fit into the active site.
3 Enzymes work best in a narrow pH range. If it varies too much then they might denature.

4.3
1 In the mitochondria.
2 Energy (from respiration). Muscles also need a stimulus.
3 Proteins.

4.4
1 Fats and oils.
2 Amylase, protease and lipase (all of them!).
3 Fatty acids and glycerol.

4.5
1 (Slightly) alkali.
2 To neutralise stomach acid.
3 Hydrochloric acid.

4.6
1 You do not need very much as it is much sweeter than glucose.
2 It makes digestion of the food easier for the baby, as some has been partly digested already.
3 It would denature the enzymes so the washing powder would no longer work.

End of chapter questions
1 The amount of energy necessary to start a reaction.
2 The area of the enzyme where the reacting substances fit in.
3 To build larger molecules from smaller ones, muscle contraction and to maintain a constant body temperature (mammals and birds only).
4 The breakdown of proteins into amino acids by the enzyme protease.
5 We use the term 'denatured'.
6 Carbon dioxide and water.
7 The pancreas, stomach and small intestine.
8 It is made by the liver and stored in the gall bladder.

Chapter 5

Pre Test
1 Carbon dioxide and water.
2 They are converted into urea and excreted.

3 In the liver.
4 The ion content of cells is linked to osmosis so a cell could take up or lose too much water if it was not controlled.
5 The thermoregulatory centre of the brain and receptors in the skin.
6 The energy needed for the sweat to evaporate comes from the surface of the skin.
7 In the sweat glands.
8 It increases the respiration rate. Some energy produced is released as heat.
9 Insulin.
10 In the pancreas.

Check yourself
5.1
1 Respiration.
2 Urea.
3 In the bladder.
5.2
1 The thermoregulatory centre of the brain and receptors in the skin.
2 In sweat glands under the surface of the skin.
3 Some of the energy released from the increased respiration is released as heat.
5.3
1 The pancreas.
2 In the liver.
3 Diabetes.

End of chapter questions
1 It affects the strength of the solution in the cell and so affects osmosis.
2 To monitor the body's temperature and then coordinate a response if the core body temperature is too high or too low.
3 The sweat evaporates from the skin's surface. The energy for this comes from the skin.
4 The pancreas detects a high sugar level, the pancreas produces insulin, some glucose is converted to glycogen and stored in the liver.
5 By radiation.
6 Through the kidneys, dissolved in water to form the urine.
7 The blood vessels constrict.
8 Osmosis.

Chapter 6

Pre Test
1 In all body cells (but not sex cells).
2 Two cells are formed.
3 For growth and to replace cells.
4 Four cells.
5 Alleles.
6 They can still differentiate into new specialised cells and so are seen as a possible cure for some diseases and conditions.
7 23 pairs.
8 Deoxyribonucleic acid.
9 An allele that 'masks' the effect of the other allele (the recessive one).
10 The nervous system.

Check yourself
6.1
1 To produce new cells for growth and to replace cells.

2 They have exactly the same genetic make up as the parent cell.
3 Two cells.
6.2
1 They can differentiate and grow into more specialised cells – this may help cure some conditions in the future.
2 Stem cells that are able to differentiate into many different specialised cells are found in developing embryos. These must be destroyed to get the stem cells.
3 Adult bone marrow contains stem cells that will only develop into a limited range of specialised cells.
6.3
1 Four sex cells.
2 Twice.
3 There is a mixing of genetic information. Half of the information comes from the male parent and the other half from the female parent.
6.4
1 Gregor Mendel.
2 Deoxyribonucleic acid.
3 Short lengths of DNA.
6.5
1 23 pairs (46).
2 XY.
3 An allele where the effect is masked by a dominant allele. An organism needs both recessive alleles if it is to show that characteristic.
6.6
1 The nervous system.
2 An organism that carries an allele (recessive) for the condition but does not have the symptoms.
3 They are looking for alleles that will result in the organism having some form of disease or disorder.

End of chapter questions
1 The stem cells develop into specialised cells.
2 Four sex cells that are all different to each other and the parent cell.
3 An allele that masks the effect of the recessive allele. An organism needs only one of these alleles in the pair to show that characteristic.
4 A short length of DNA that controls one characteristic.
5 They make a copy of themselves.
6 23 pairs.
7 Pairs of genes controlling the same characteristic.
8 A disorder of cell membranes.

EXAMINATION-STYLE QUESTIONS

1 (a) (i) The lungs.
        (ii) The liver.
        (iii) The kidney.
        (iv) The skin.   (4 marks)
   (b) As it evaporates, it takes energy from the skin, cooling the skin down.   (3 marks)
   (c) Any two of: muscles contract, they respire to produce the energy necessary, some energy is lost as heat.   (2 marks)

2 (a) In developing embryos and adult bone marrow.   (2 marks)
   (b) Any two of: stem cells can differentiate (develop), into different types of cell, possible cure for certain conditions / diseases.   (2 marks)
   (c) The most useful stem cells come from developing embryos, you destroy the embryo by taking the stem cells, some people argue, therefore, that you have destroyed a 'life'.   (3 marks)
3 (a) (i) 2   (1 mark)
        (ii) By carrying out the readings more than once to avoid anomalous results (it does state all other variables were kept the same).   (1 mark)
        (iii) By using a narrower band of pH readings with smaller intervals between them, e.g. 1.2, 1.4, 1.6 etc.   (2 marks)
   (b) In the stomach, the only part where protease works in acid conditions.   (2 marks)
   (c) (i) Lipase digest fats, into fatty acids, and glycerol. Amylase digests starch, into glucose (sugar).   (5 marks)
        (ii) Two of: In the mouth (salivary glands), pancreas, small intestine.   (2 marks)
   (d) To neutralise, acid from the stomach.   (2 marks)
4 (a) It is the energy source for cells.   (1 mark)
   (b) The pancreas.   (1 mark)
   (c) The pancreas produces insulin.   (1 mark)
        Two of: excess glucose is converted into glycogen, by the liver, glycogen is stored in the liver (and some muscles).   (2 marks)
   (d) Either through eating an appropriate diet, or through insulin injections.   (2 marks)
5 (a) Diagram should show that mother is Hh and father is hh.
        D has inherited the dominant H allele from his mother and the recessive h allele from his father, and therefore has the disease. (Note that mother cannot be HH or all of her children would have the disease.)   (3 marks)
   (b) Inherited the recessive gene from the father, and the recessive gene from the mother.   (2 marks)
   (c) 50%
        The recessive from parent F has a 1 in 2 chance of pairing with the dominant or the recessive gene from the person with H.   (2 marks)
   (d) (i) Two of: It codes for amino acids, these link to form proteins, proteins control the characteristics we have.   (2 marks)
        (ii) A gene.   (1 mark)

Chapter 1

Pre Test

1 The substances are taken up against a concentration gradient, so neither diffusion nor osmosis would take place in the right direction.
2 It requires energy produced through respiration.
3 In mitochondria.
4 Alveoli or air sacs.
5 Diffusion takes place more easily through moist surfaces.
6 Capillaries.
7 Small soluble molecules of digested food [alternatively name the products e.g. glucose, amino acids, fatty acids and glycerol].
8 Stomata (singular: stoma).
9 Carbon dioxide (waste product of respiration).
10 One of: higher temperature, more wind, drier air (less humidity).

Check yourself

1.1
1 Molecules randomly move from areas of higher concentration to areas of lower concentration.
2 The difference in concentration between the two areas where the substance is present.
3 The molecules have to be actively moved by the cell – diffusion and osmosis would work in the opposite direction. This active movement requires energy.

1.2
1 Carbon dioxide.
2 To help diffusion take place more quickly, as there is a greater surface over which it can take place.
3 Capillaries.

1.3
1 In the small intestine.
2 Villi (singular: villus).
3 Diffusion and active transport.

1.4
1 To increase the area over which gases can exchange and, therefore, speed up the process.
2 Gases diffuse much more easily through moist surfaces.
3 So that the net movement of the molecules will continue at a high rate – diffusion will be faster.

1.5
1 Oxygen.
2 Stomata (singular: stoma).
3 Root hairs.

1.6
1 To allow the exchange of gases for photosynthesis and respiration.
2 Guard cells.
3 Water is less likely to evaporate from them, they are not in the direct Sun and are protected from the wind to some extent.

End of chapter questions

1 The particles are moving from an area of lower concentration to an area of higher concentration.
2 Respiration.
3 There is less distance across which diffusion has to take place.
4 So that the digested food can be taken away quickly and efficiently.
5 They have large surface areas.
6 There is less distance inside the leaf over which substances have to diffuse.
7 The air already is holding quite a lot of water – it is more difficult to take more from the plants.
8 Water and mineral ions.

Chapter 2

Pre Test

1 Capillaries.
2 The kidney.
3 Haemoglobin.
4 They need more oxygen and glucose and they need to get rid of more carbon dioxide.
5 Lactic acid.
6 The need to continue to break down any remaining lactic acid after exercise has been completed – this requires oxygen.
7 Glucose, most of the water, some of the mineral ions.
8 They may need to be reabsorbed against the concentration gradient.
9 The immune system.
10 Dialysis.

Check yourself

2.1
1 It has muscular walls, which can contract.
2 Veins.
3 Carbon dioxide.

2.2
1 The kidneys.
2 At all living (respiring) cells.
3 So that more haemoglobin is able to 'pack' into them.

2.3
1 So that more oxygen can be taken into the lungs and more carbon dioxide can be breathed out from them.
2 So that more blood can flow to the lungs carrying oxygen and glucose to them (and carbon dioxide away) – liquids flow more easily through wider vessels.
3 Glycogen.

2.4
1 A lack of oxygen.
2 A build up of lactic acid.
3 Carbon dioxide and water.

2.5
1 Urea is not reabsorbed at all.
2 It may be reabsorbed against the concentration gradient.
3 More on a cold day than on a warm day, as more water is lost by sweating to keep cool.

2.6
1 Glucose, most of the water and some of the mineral ions.
2 A partially permeable membrane.
3 The composition of the blood is changing all of the time, it needs to be filtered so that toxins (poisons) do not build up.

2.7
1 The white blood cells would attack the kidney, as they would not recognise it as part of the body.
2 The greater the difference, the more readily the body's immune system will try to destroy it.
3 People that have just died (as long as the kidneys were not the reason for the person dying).

End of chapter questions

1 Arteries.
2 Combined with haemoglobin to form oxy-haemoglobin.
3 Aerobic respiration.
4 To pump more blood, carrying oxygen and glucose, to actively respiring muscles and to take away the increased amount of carbon dioxide.
5 Lactic acid.
6 The build-up of lactic acid during anaerobic respiration.
7 Urea, mineral ions and water.
8 The concentration of glucose in the blood and in the dialysis fluid is the same, so there is no net diffusion.

EXAMINATION-STYLE QUESTIONS

1 (a) Any two of: thin walls; large surface area; moist.   (2 marks)
   (b) Combined with haemoglobin, to form oxy-haemoglobin.   (2 marks)
   (c) (i) Villi (one is a villus)   (1 mark)
       (ii) (Random) movement of particles from an area of high concentration to an area of lower concentration, the process requires no energy from the body.   (3 marks)
       (iii) Movement of particles against the concentration gradient requires energy (from respiration).   (2 marks)
2 (a) (i) Guard cells.   (1 mark)
       (ii) To open and close the stomata.   (1 mark)
   (b) One of: it is very thin; it contains a lot of air spaces.   (1 mark)

(c) Carbon dioxide, the plant would be photosynthesising (more rapidly than respiring). (2 marks)
(d) (i) Transpiration. (1 mark)
(ii) Any two of: warmer temperature – the air holds more water or water molecules have more energy to evaporate; less humid – the air will hold more water evaporating from the plant; more windy – any water in the air around the plant will be blown away so more water can evaporate. (4 marks)
3 (a) Artery. (1 mark)
(b) (i) Any two of: carbon dioxide; products of digestion; urea. (2 marks)
(ii) Any two of: carbon dioxide – living cells/organs → lungs; products of digestion – small intestine → living cells/organs; urea – liver → kidneys. (4 marks)
(c) (i) Any two of: The (muscle) cells need more energy. More oxygen is brought to them. More glucose is brought to them. More carbon dioxide is taken away. (3 marks)
(ii) Less energy is released. Lactic acid is produced. The runner's muscles become fatigued (tired). (3 marks)
(iii) They still have lactic acid. This needs to be broken down. They still need to breathe in more oxygen to do this. (3 marks)
4 (a) Filtration. (1 mark)
(b) (i) Glucose (sugar). (1 mark)
(ii) Water or mineral ions. (1 mark)
(c) (i) Partially permeable. (1 mark)
(ii) So that these substances don't diffuse out of the blood, if they did they would need to be reabsorbed. (2 marks)
(d) Any two of: There are not that many kidney donors / not enough kidneys. The donor kidney has to be a very good tissue match. Some transplants fail, putting up the overall cost. Surgery can be dangerous, especially for some people. (2 marks)

Chapter 3

Pre Test
1 Bacteria are used to make yoghurt and cheese.
2 Yeast is used to make wine, beer and bread.
3 Ethanol and carbon dioxide.

4 Sugars from the barley grains.
5 Lactose sugar.
6 *Penicillium*.
7 A vessel that produces biogas (mainly methane) from the breakdown of waste material and plant material.
8 Ethanol (alcohol).
9 It is a jelly-like substance that contains nutrients necessary for microorganisms to grow.
10 At that temperature harmful microorganisms (pathogens) could grow.

Check yourself
3.1
1 Carbohydrate.
2 Pathogens.
3 Inoculating loop.
3.2
1 Barley grains.
2 Hops.
3 Enzymes.
3.3
1 Yoghurt and cheese.
2 Anaerobic respiration.
3 Lactose.
3.4
1 pH and temperature.
2 If it became too hot the enzymes in the microorganisms could denature.
3 It is a good source of protein and can be eaten by vegetarians.
3.5
1 *Penicillium*.
2 As a source of energy for the mould.
3 Nitrogen.
3.6
1 Methane.
2 Waste or plant material.
3 Respiring anaerobically.
3.7
1 Carbohydrase enzymes.
2 Anaerobically.
3 It must be distilled.

End of chapter questions
1 A Petri dish.
2 They are passed through a flame.
3 To prevent microorganisms getting in or out.
4 Ethanol (alcohol) and carbon dioxide.
5 Yeast needs to be able to grow and reproduce quickly, aerobic respiration produces more energy for it to be able to do this.
6 If the pH changes too much the enzymes will not work so effectively and could denature (be destroyed), so the whole process would stop.
7 Starch.
8 Once the nitrogen is mostly used up.

EXAMINATION-STYLE QUESTIONS

1 (a) The reactions will take place faster. Enzymes work better at warmer temperatures. (2 marks)

(b) (i) Lactose sugar.
(ii) Lactic acid.
(iii) The lactic acid causes the milk to clot. (3 marks)
2 (a) (i) To provide oxygen for aerobic respiration of bacteria. (2 marks)
(ii) Bacteria respiring release some energy as heat. (2 marks)
(iii) pH. (1 mark)
(b) (i) *Penicillium*. (1 mark)
(ii) Sugars and nitrogen. (2 marks)
(iii) Once most of the nitrogen is used up. (1 mark)
(c) (i) He was trying to prove that once any organisms were killed (by the boiling) they couldn't just reappear if the glass was sealed. (1 mark)
(ii) Wash their hands between working with patients. (1 mark)
(iii) So that any bacteria / germs were washed away or killed and didn't infect the next patient. (2 marks)
(iv) Any two of: Doctors didn't like to admit they may have been killing patients. Washing hands at the time was a strange thing to do. No-one could actually see these microorganisms. (2 marks)
3 (a) They allow them to start germinating, so starches are broken down into sugars. (2 marks)
(b) (i) So that there is enough energy for it to grow and reproduce quickly. (1 mark)
(ii) Anaerobic respiration or (anaerobic) fermentation. (1 mark)
(iii) Ethanol. (1 mark)
(iv) Carbon dioxide. (1 mark)
(c) Hops. (1 mark)
(d) (Natural) sugars. (1 mark)
4 (a) Methane. (1 mark)
(b) (i) Plant material, waste material from animals or plants. (2 marks)
(ii) Anaerobic respiration. (2 marks)
(c) Any two of: There may be no other energy source. They may not be able to pay for their energy. They may have plenty of waste animal and plant material. (2 marks)
(d) Possibly remote areas where it is expensive to get other energy sources to; possibly where there is plenty of plant and waste plant material available locally. (2 marks)